CORROSION OF METALS UNDER THERMAL INSULATION

A symposium sponsored by
ASTM Committees C-16 on Thermal
Insulation and G-1 on Corrosion
and the National Association of
Corrosion Engineers, the Institution
of Corrosion Science and Technology, and
the Materials Technology Institute
of the Chemical Process Industries

ASTM SPECIAL TECHNICAL PUBLICATION 880
Warren I. Pollock, E. I. du Pont de Nemours
and Company, and Jack M. Barnhart,
Thermal Insulation Manufacturers
Association, editors

ASTM Publication Code Number (PCN)
04-880000-27

1916 Race Street, Philadelphia, PA 19103

Library of Congress Cataloging in Publication Data

Main entry under title:

Corrosion of metals under thermal insulation.

(ASTM special technical publication; 880)
Papers presented at the symposium held at San Antonio, TX, 11-13 Oct. 1983.
Includes bibliographies and index.
"ASTM publication code number (PCN) 04-880000-27"
1. Corrosion and anti-corrosives—Congresses.
2. Insulation (Heat)—Congresses. I. Pollock, Warren I. II. Barnhart, Jack M. III. ASTM Committee C-16 on Thermal Insulation. IV. Series.
TA462.C6567 1985 621.1′623 85-10616
ISBN 0-8031-0416-2

NOTE

The Society is not responsible, as a body,
for the statements and opinions
advanced in this publication.

Printed in Ann Arbor, MI
Aug. 1985

Foreword

The symposium on Corrosion of Metals Under Thermal Insulation was presented at San Antonio, TX, 11–13 Oct. 1983. The symposium was sponsored by ASTM Committees C-16 on Thermal Insulation and G-1 on Corrosion and by the National Association of Corrosion Engineers, the Institution of Corrosion Science and Technology, and The Materials Technology Institute of the Chemical Process Industries. Warren I. Pollock, E. I. du Pont de Nemours and Company, and Jack M. Barnhart, Thermal Insulation Manufacturers Association, presided as chairmen of the symposium and are editors of this publication.

Related ASTM Publications

Atmospheric Corrosion of Metals, STP 767 (1982), 04-767000-27

Atmospheric Factors Affecting the Corrosion of Engineering Metals, STP 646 (1978), 04-646000-27

Chloride Corrosion of Steel in Concrete, STP 629 (1977), 04-629000-27

A Note of Appreciation to Reviewers

The quality of the papers that appear in this publication reflects not only the obvious efforts of the authors but also the unheralded, though essential, work of the reviewers. On behalf of ASTM we acknowledge with appreciation their dedication to high professional standards and their sacrifice of time and effort.

ASTM Committee on Publications

ASTM Editorial Staff

Contents

Introduction

Very serious corrosion problems can occur to plant equipment, tankage, and piping components that are thermally insulated if the insulation becomes wet. Many companies have had to repair or replace major pieces of equipment at considerable expense. At one chemical process plant alone, the cost was reported to be in the millions of dollars.

On carbon steels, the corrosion is usually of a general or pitting type. On austenitic stainless steels, the corrosion is almost always chloride stress corrosion cracking. It is an insidious problem. The insulation usually hides the corroding metal and the problem can go undetected for years until metal failure occurs. This sometimes occurs five or more years after the insulation becomes wet.

Insulation materials received from manufacturers and distributors are dry, or nearly so. Obviously, if they remain dry there is no corrosion problem. So, the solution to the corrosion under wet insulation problem would appear to be fairly obvious: keep the insulation dry or protect the metal.

Unfortunately, application of these solutions is not that simple. Insulation can get wet in storage and field erection. Weather barriers are not always installed correctly or they are not effective in fully preventing water ingress. Weather barriers and protective coatings get damaged and are not maintained and repaired.

To further complicate the problem, it appears that the degree of corrosion when an insulation gets wet is dependent on the type of insulation. Some insulations contain elements that promote corrosion, such as chloride stress corrosion cracking of austenitic stainless steels.

Inspection for the problem is often difficult. Good inspection techniques to determine that the insulation is wet or that the metal surface is corroded or stress cracked have not been widely available.

Many companies have developed the practice of applying a protective coating to steels to keep moisture from contacting the metal. Some do this only for carbon steels, some only for stainless steels, some for both. What coatings to use have varied considerably from one plant site to another.

Wet insulation is significantly less thermally efficient than dry insulation. This alone should be a high driving force for keeping insulation dry, but, interestingly, this has not been the case on many plant sites.

Little has appeared on this overall problem in the literature, and there has not been a major conference in North America before this one. In Nov. 1980, a conference was held in Britain on "Corrosion Under Lagging." The success of that meeting stimulated the organization of a similar type conference in the U.S.

The purpose of this conference was to provide a forum for a thorough review of the problem and the various control and inspection methods being used and under development. Because the problem is broad based, several technical societies were cosponsors: ASTM Committee C-16 on Thermal Insulation and Committee G-1 on Corrosion; the National Association of Corrosion Engineers (NACE); Materials Technology Institute of the Chemical Process Industries (MTI); and Institution of Corrosion Science and Technology, a sponsor of the British conference.

The conference was very successful with some 150 people attending. It provided high recognition to a costly problem where the solutions are many faceted, as indicated by the papers in this publication.

Warren I. Pollock

E. I. du Pont de Nemours & Co., Inc., Engineering Department, Wilmington, DE 19898; symposium cochairman and editor.

Technical Overview

Jack M. Barnhart[1]

The Function of Thermal Insulation

REFERENCE: Barnhart, J. M., **"The Function of Thermal Insulation,"** *Corrosion of Metals Under Thermal Insulation, ASTM STP 880,* W. I. Pollock and J. M. Barnhart, Eds., American Society for Testing and Materials, Philadelphia, 1985, pp. 5–8.

KEY WORDS: corrosion, insulation, energy, chlorides, stress corrosion

Thermal insulations or thermal insulation systems are usually defined as materials or combinations of materials that retard the flow of heat energy by conductive, convective, or radiative modes of transfer or a combination of these. In order to be effective, they must be properly applied.

Primarily, thermal insulation serves one or more of the following functions:

1. Conserve energy by reducing heat loss or gain of piping, ducts, vessels, equipment, and structures.
2. Control surface temperatures of equipment and structures for personnel protection and comfort.
3. Facilitate temperature control of a chemical process, a piece of equipment, or a structure.
4. Prevent vapor condensation at surfaces having a temperature below the dew point of the surrounding atmosphere.
5. Reduce temperature fluctuations within an enclosure when heating or cooling is not needed or available.

Thermal insulations may also serve additional functions:

1. Add structural strength to a wall, ceiling, or floor section.
2. Provide support for a surface finish.
3. Impede water vapor transmission and air infiltration.
4. Prevent or reduce damage to equipment and structures from exposure to fire and freezing conditions.
5. Reduce noise and vibration.

[1]Thermal Insulation Manufacturers Association, 7 Kirby Plaza, Mt. Kisco, NY 10549.

Thermal insulation is used to control heat flow in temperature ranges from near absolute zero through 1650°C (3000°F) and higher. Insulations normally consist of the following basic materials and composites:

(1) inorganic, fibrous or cellular materials such as glass, asbestos, rock or slag wool, calcium silicate, bonded perlite, vermiculite, and ceramic products.

(2) organic fibrous materials, such as cotton, animal hair, wood, pulp, cane, or synthetic fibers, and organic cellular materials, such as cork, foamed rubber, polystyrene, polyurethane, and other polymers.

(3) metallic or metalized organic reflective membranes (which must face air, gas-filled, or evacuated spaces).

The structure of mass-type insulation may be cellular, granular, or fibrous, providing gas-filled voids within the solid material that retard heat flow. Reflective insulation consists of spaced, smooth-surfaced sheets made of metal foil or foil surfaced material that derives its insulating value from a number of reflective surfaces separated by air spaces.

The physical forms of industrial and building insulations are

(1) loose fill and cement,
(2) flexible and semirigid,
(3) rigid,
(4) reflective, and
(5) foamed in place.

Depending on design requirement, the choice of a particular thermal insulation may involve a set of secondary characteristics in addition to the primary property of low-thermal conductivity. Characteristics, such as resiliency or rigidity, acoustical energy absorption, water vapor permeability, air flow resistance, fire hazard and fire resistance, ease of application, applied cost, or other parameters, may influence the choice among materials having almost equal thermal performance values.

Some insulations have sufficient structural strength for use as load-bearing materials. They may be used occasionally to support load-bearing floors, form self-supporting partitions, or stiffen structural panels. For such applications, one or more of the following properties of an insulation may be important: strength in compression, tension, shear, impact, and flexure. These temperature dependent mechanical properties vary with basic composition, density, cell size, fiber diameter and orientation, type and amount of binder (if any), and both temperature and environmental conditioning.

The presence of water as a vapor, liquified or solid in insulation will decrease its insulating value; it may cause deterioration of the insulation and eventual structural damage by rot, corrosion, or the expansion action of freezing water. Whether or not moisture accumulates within the insulation depends on the hygroscopic properties of the insulation, operating temperatures, ambient conditions, and the effectiveness of water vapor retarders in relation to other vapor resistances within the composite structure.

The moisture resistance depends on the basic material of the insulation and the type of physical structure. Most insulations are hygroscopic and will gain or lose moisture in proportion to the relative humidity of the air in contact with the insulation. Fibrous and granular insulations permit transmission of water vapor to the colder side of the structure. A vapor retarder should therefore be used with these materials where moisture transmission is a factor. Other insulations having a closed cellular structure are relatively impervious to water and water vapor. Properties that express the influence of moisture include: absorption (capillarity), adsorption (hygroscopicity), and the water vapor transmission rate.

Other properties of insulating materials that may be important, depending upon the application, include: density; resilience; resistance to settling; permanence; reuse or salvage value; ease of handling; dimensional uniformity and stability; resistance to chemical action and chemical change; ease in fabricating, applying, or finishing; and sizes and thickness obtainable.

In some specific applications, thermal insulation is called on to perform another function, namely, to retard chloride induced stress corrosion.

An inherent characteristic of austenitic stainless steel is its tendency to crack at stress points when exposed to certain corrosive environments. The mechanisms of stress corrosion cracking are complex and incompletely understood, but apparently related to certain metallurgical properties. Chloride ions concentrated at a stress point will catalyze crack propagation. Other halide ions are also suspect.

Chlorides are common to most environments, so great care must be taken to protect austenitic stainless steels. Water, dust and soil, process liquids, chemical fumes, even the air in coastal regions, contain chlorides in measurable, and thus additively significant quantities.

Most thermal insulations will not, in themselves, cause stress corrosion cracking as may be shown by tests. However, when exposed to environments containing both chlorides and moisture, insulation systems may act as collecting media, transmigrating and concentrating chlorides on heated stainless steel surfaces. If, however, moisture is not present, the chloride salts cannot migrate, and stress corrosion cracking will not take place.

Insulations may also be specially formulated to inhibit stress corrosion cracking in the presence of chlorides through modifications in basic composition or incorporation of certain chemical additives. Stress corrosion cracking is a metallurgical shortcoming of austenitic stainless steel. It is unrealistic to expect an insulation to overcome this shortcoming. If the conditions are such that stress corrosion cracking will occur, then, the very best an insulation could hope to do is delay the inevitable. This is demonstrated by the occurrence of stress corrosion cracking under insulations that were mostly sodium silicate, a known "inhibitor." Stress corrosion cracking under insulations is not a simple insulation problem; to quote from William G. Ashbaugh of Union Carbide in a paper presented to the National Association of Corrosion Engineers:

> The inhibition of insulation by the addition of neutralizers or other agents to the insulation is insufficient protection against externally introduced chlorides which are the major source of stress corrosion cracking.

and from later in the paper:

> The author does not claim that insulation materials cannot or will never cause stress corrosion cracking, but plant experience and laboratory screening tests indicate that most insulation materials which remain relatively dry play only a secondary role in stress corrosion cracking. The real problem in chemical plants exists as a result of the combination of corrosive atmosphere and the many types of crevices, joints, and areas where atmospheric chloride contamination and concentration can occur.

The real problem and need of insulation manufacturers is to inform the users of the "real problem" and ways to address it. The insulation cannot and should not be forced to overcome the shortcomings of austenitic stainless steel when used in real world environments.

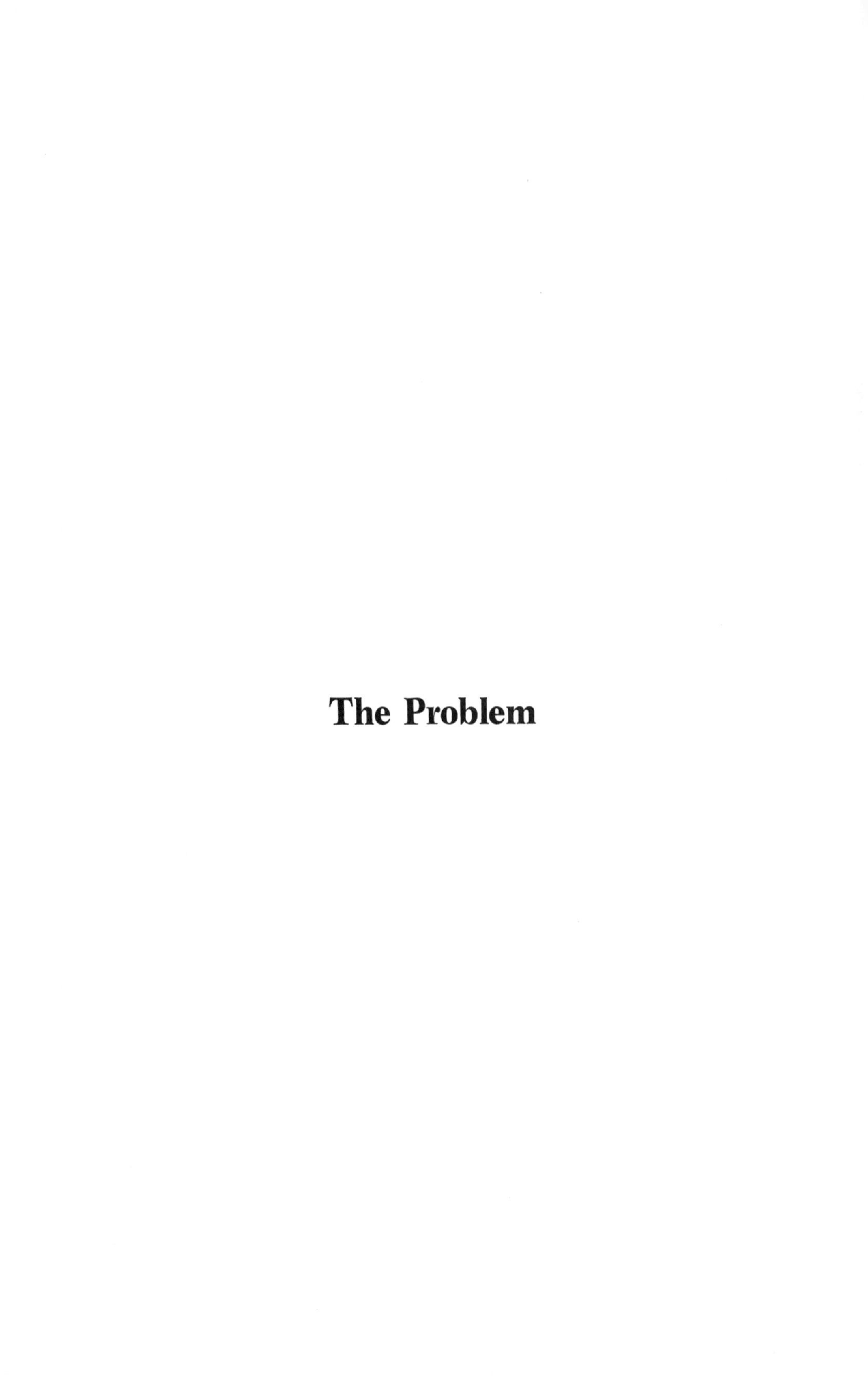

The Problem

Peter Lazar, III[1]

Factors Affecting Corrosion of Carbon Steel Under Thermal Insulation

REFERENCE: Lazar, P., III, **"Factors Affecting Corrosion of Carbon Steel Under Thermal Insulation,"** *Corrosion of Metals Under Thermal Insulation, ASTM STP 880*, W. I. Pollock and J. M. Barnhart, Eds., American Society for Testing and Materials, Philadelphia, 1985, pp. 11-26.

ABSTRACT: The problem of corrosion under insulation is studied by analysis of seven contributing factors: (1) equipment design; (2) service temperatures; (3) insulation selection; (4) paints and coatings; (5) weather barriers; (6) climate; and (7) maintenance practices. The purposes of our study are to identify optimum locations for inspection of insulated equipment and to identify improvements in new equipment design and insulation, as well as maintenance of existing systems. Emphasis is placed on equipment design features, operating temperature, and weather barrier deficiencies to guide inspection. Weather barrier maintenance is critical for prevention of corrosion on existing equipment. New insulated equipment designs should provide special details so that the system can be better sealed against water entry and allow clearance for the specified insulation thickness. On new equipment or when reinsulating existing equipment, a high quality coating system is recommended, since touch up opportunity does not exist.

KEY WORDS: corrosion, insulation, paints, coatings, maintenance, inspection, temperature, climate, petrochemical equipment, mechanical design, weather barriers

The amount of carbon steel lost because of corrosion under insulation (CUI) is determined by (1) wet exposure cycle characteristics (duration and frequency), (2) corrosivity of the aqueous environment, and (3) failure of protective barriers (paint and jacketing). There are numerous controllable factors in the design, construction, and maintenance of insulated equipment that affect the variables just mentioned, and therefore the amount of damage caused by CUI. In our studies of CUI problems in our plants, seven controllable factors were identified. They are

[1]Staff engineer, Exxon Chemical Americas, P.O. Box 241, Baton Rouge, LA 70821.

(1) equipment design,
(2) service temperatures,
(3) insulation selection,
(4) protective paints and coatings,
(5) weather barriers,
(6) climate, and
(7) maintenance practices.

These factors will be examined individually to demonstrate how past common practices supplied the requirements for corrosion. Understanding how common practices cause the conditions for CUI is leading to better inspection of existing equipment and better design of new equipment.

Equipment Design

The design of pressure vessels, tanks, and piping generally includes numerous details for support, reinforcement, and connection to other equipment. These details include stiffening rings, gussets, brackets, reinforcing pads, flanges, and so forth. Design of equipment, including these details, is the responsibility of engineers or designers who use construction codes to assure consistently reliable designs for both insulated and noninsulated subjects. Consideration of the problem of insulating those details and of leaving room for the insulation is lacking in those codes and in the instructions to the designers; thus, the equipment is designed like those that would not be insulated. The weather barrier on such designs is broken frequently because of inappropriate details for insulated equipment or the lack of space for the specified thickness of insulation.

The consequence of broken jacketing is that more water gets into the insulation at each exposure cycle, taking longer to dry, cooling the insulated equipment item to temperatures where corrosion is possible, and increasing the amount of cumulative damage. Some of the equipment details, such as gussets, actually channel water into the insulation. There are also economic consequences such as energy inefficiency and construction costs. The inefficiency of wet insulation is obvious. Also obvious is the cost of insulating equipment not designed for insulation, as one watches insulators cut up insulation and jacketing and sees needless hours spent installing around complicated details.

The solution to the factor of equipment design affecting CUI is to take an integrated approach to the design. Specify the insulation thickness and type, and the jacketing type before designing the equipment. Define acceptable "code" details for the weatherproofing type, and specify spacing standards. In every case, simplify the surface to be insulated.

See Figs. 1 and 2.

FIG. 1—*Opening in metal jacketing, cut by insulators to accommodate piping that was run too close to a pressure vessel for the specified thickness. Piping has been moved as part of our effort to rectify deficiencies.*

Service Temperatures

Service temperature is important in CUI for two reasons:

(1) higher temperatures allow water to be present against the steel for less time, but
(2) higher temperatures make the water more corrosive, and paints and caulking fail sooner.

Generally, equipment that operates below freezing a large fraction of its life is protected against corrosion; however, attachments to that equipment, which are not as cold, are vulnerable in the transition out of the vapor barrier into warm humid air. For the most part, corrosion associated with equipment oper-

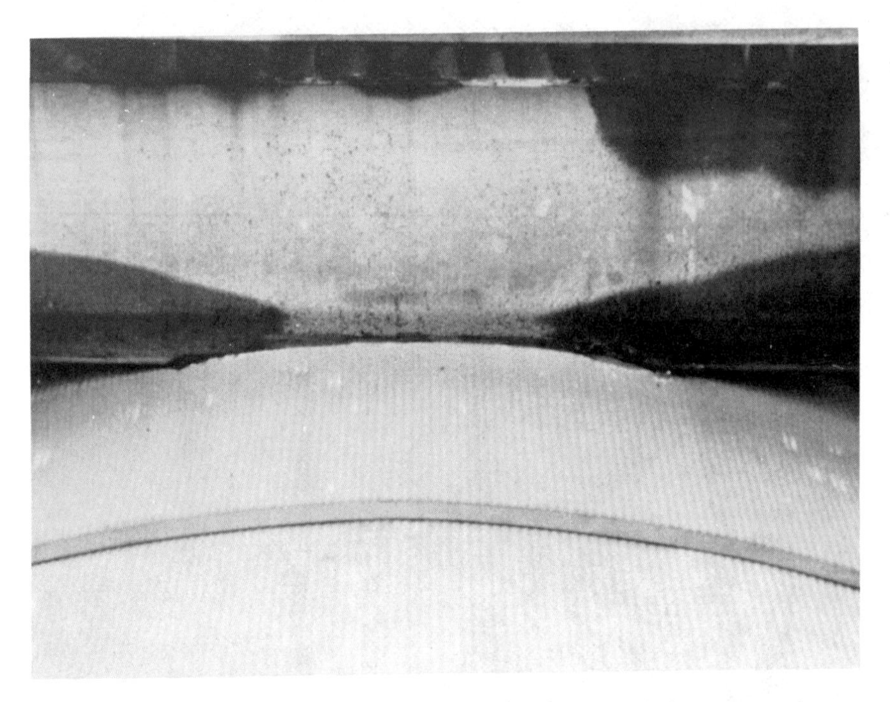

FIG. 2—*Jacketing cut by insulators to accommodate I-beam for platform support that was closer to the vessel than the insulation thickness specified. Water can run on the flange and enter the insulation at this point.*

ating below freezing temperatures is corrosion outside of, not under insulation. Corrosion of equipment operating between freezing and atmospheric dew points suffers less localized corrosion, and corrosion rates tend to be lower because, first, the water temperature is lower and second, because contaminants are continuously diluted by condensation; however, since the corrosion occurs continuously, damage can accumulate almost as quickly as it does under warm insulation.

Corrosion under warm insulation is more difficult to manage or understand because of the dryout of entering water. Dryout produces surprisingly corrosive conditions on a cyclic basis, as well as less than adequate performance by many protective coatings on which we often rely. The following is a summary of some of our observations vis-a-vis warm service.

1. The temperature range of 60 to 80°C appears to account for the greatest amount of damage, but failures have occurred even on systems operating at or above 370°C, when weatherproofing is poorly maintained.

2. On very warm equipment with relatively small weatherproofing defects, corrosion will tend to be at points of entry of water where rapid evaporation occurs. As equipment temperatures are reduced or weatherproofing defects get

larger, water is allowed to run to lower points where it is held up to dry more slowly, or not at all.

3. Annual corrosion damage rates may exceed 1.5 $mm \cdot y^{-1}$. The corrosivity is partly a function of water temperature itself, but also a function of concentration of salts carried in with the water, drying out in the same locations repeatedly.

The temperature on some equipment varies by location, especially on towers. For example, temperatures can range from more than 80°C on the bottom to less than 0°C at the top. This produces extremes of corrosion condition on a single equipment item.

See Figs. 3 through 8.

Insulation Selection

Insulation characteristics most influential on corrosion under insulation are water absorbancy and chemical contributions to the water phase. While no insulation selection will preclude the possibility of corrosion, some insulation types leave the system less sensitive to defects in weatherproofing or paint film, because they are nonabsorbent and chemically benign.

FIG. 3—*Corrosion above an insulation support ring and around a small pipe connection. This is a vertical drum that had been heated with a steam coil at one time. Insulation rings act as a hold up for water entering through deficient top head weatherproofing.*

FIG. 4—*Insulation for personnel protection has in the past been stopped about 2 m above grade. Severe corrosion at the open end of the insulation system (water entry point) is typical of very hot systems such as steam lines.*

Unfortunately, insulation selection has not been based on any consideration of maintenance costs; rather, it has been based on installed cost versus energy cost saved.

Other considerations, which are normally neglected, include:

(1) repairability of the insulation. Some has to be removed for inspection, periodically, while some is accidentally damaged;

(2) effect of absorbency on steel corrosion costs and paint film life; and

(3) credit for cyclic service energy savings on behalf of nonabsorbent insulations, because of less water needing to be boiled away.

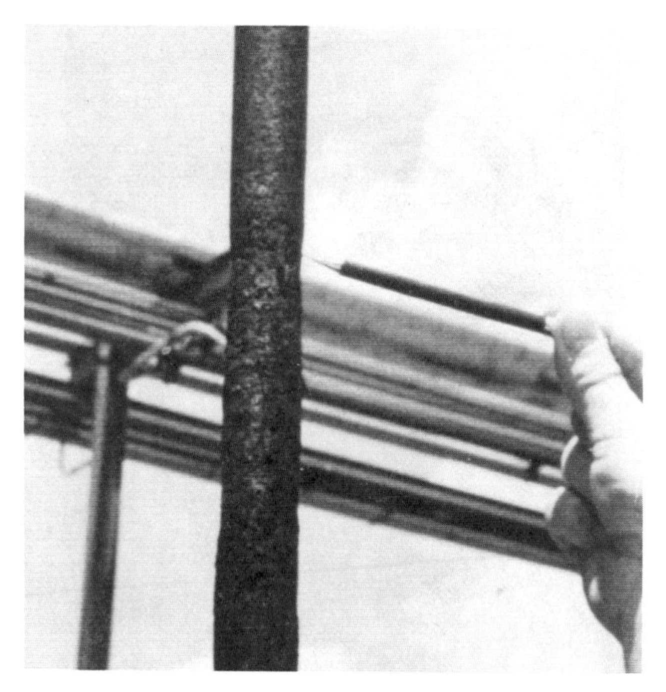

FIG. 5—*Another example of the effect of corrosion under personnel protection. Note the reduction in diameter of the pipe where indicated, corresponding to the water entry point.*

These kinds of considerations are virtually impossible to model for cash flow analysis, but they should not be ignored. It is therefore necessary to exercise judgment when selecting insulation, beyond acceptance of calculated returns on investments.

Cellular glass has been widely adapted by our plants for use from 150°C down, including low temperature requirements. The main advantage is zero water absorption and reasonable installation cost. Drawbacks are that the material is somewhat prone to breakage, and with rising temperatures, has an unacceptable k factor. Theoretically the hydrogen sulfide contained in the cells would contribute to corrosion when water was present between insulation and steel, but it is not released from those cells unless they are broken.

Calcium silicate insulation is highly water absorbent, and as such has contributed to much of our corrosion problems at moderate temperatures and on cyclic services. Some calcium silicate still in service may also have contained corrosive salts, although this may be corrected for the most part with new material being produced. The advantages of calcium silicate is primarily in k factor at elevated temperature versus most block insulation types. To realize this advantage requires that weatherproofing be in good condition and that the system should be in steady hot service to keep the insulation dry.

FIG. 6—*A steam traced line showing severe corrosion under the tracing tubing. Galvanic corrosion may be a factor between copper tubing and steel pipe, when water frequently enters the system.*

Although extensively used in the past, polyurethane foam (PUF) is not a popular insulation type in our plants at this time for moderate or cold services. Reasons for this are

(1) vulnerability to damage,
(2) utter dependence on the vapor barrier because of its high level of water absorbency.
(3) corrosivity of water because of hydrolysis of halogenated flame retardants needed to make the insulation safe in the plant, and
(4) sensitivity to humidity during application.

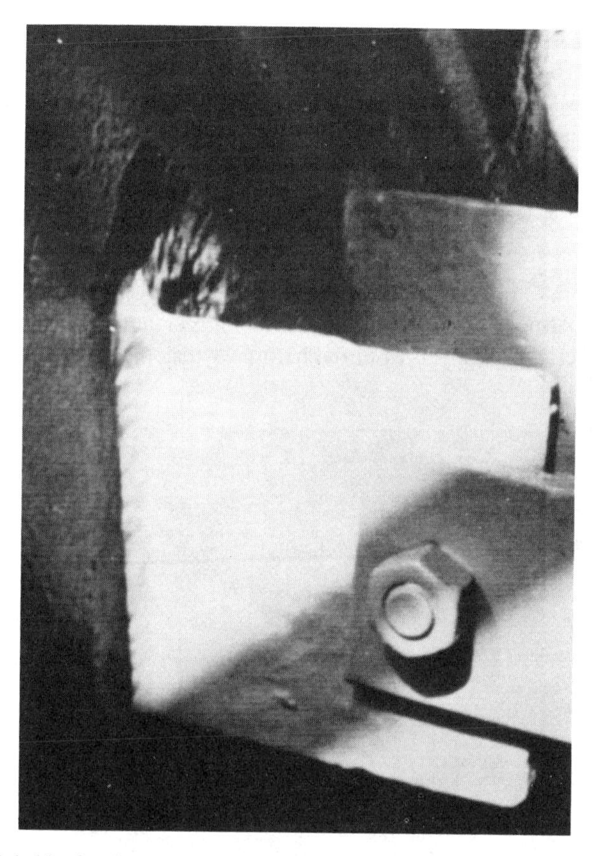

FIG. 7—*A ladder bracket on a tower operated at about 80°C. Corrosion at this water entry point is highly concentrated. Surface was abrasivly cleaned to remove scale from the corrosion trench around the bracket.*

Although a low *k* factor is claimed as an advantage for PUF, in practice, humidity during application and water entry during service often degrades this performance. The main attraction to PUF is low installation cost.

Fiberglass and mineral fiber based insulations are used judiciously in our plant, primarily where existing equipment is spaced such that only fiberous insulations will fit and do the job. Water absorption is a concern with fiberous insulation, although the absorbancy of these insulations may vary greatly from product to product.

Summarizing insulation selection and its effect on CUI, the two critical factors are absorbency, because of the effect on the amount of time required to dry out, and on wicking tendencies, and chemical contributions to an entering water phase, which increases its corrosivity in most cases. It is important to reemphasize two very important points:

1. Corrosion is possible under all types of insulation. The insulation type is only a contributing factor.

2. Insulation selection requires consideration of a large set of advantages and disadvantages in areas of installation and operating economics as well as corrosion and is by no means a simple decision.

See Fig. 9.

Protective Coatings

Protective coatings, or paint, are extremely important in preventing CUI; failure of protective coatings is essential before corrosion can occur. In the past

FIG. 8—*General corrosion under a manway on a vessel with a moderate operating temperature. General corrosion is characteristic of high water entry capacity compared to heat available for drying.*

FIG. 9—*Corrosion under PUF on an idle unit.*

the prevailing attitude has been, that a single coat of primer is adequate, on the assumption that the weatherproofing never let water into the insulation system. Consider the nature of the service in which a coating under insulation serves. From that review some direction can be taken on coating selection.

First, the service is virtually an immersion service. In general, the insulation environment is wet longer than that on the surface of most uninsulated equipment, once the weather or vaporproofing is breached. Second, under warm insulation the coating is obviously subject to higher temperatures than most painted uninsulated equipment. Consideration must be given to both chemical degradation and permeability of the coating. Highly permeable coatings allow corrosion to start behind the coating, even in the absence of breaks or pinholes. Finally, many coatings depend on some form of sacrificial inhibitor or are essentially only that (for example inorganic zinc rich coatings). Zinc rich coatings have given extremely poor performance in our plants under insulation. The following are possible explanations for that performance:

1. There is the possibility of reversal in the polarity of galvanic couples, with increasing temperature.
2. Salts carried in and deposited with the water interfere with or destroy the effectiveness of the inhibitors.
3. The subinsulation environment is not freely ventilated and may not have adequate oxygen or carbon dioxide for film forming reactions to occur.

In general, our plants prefer a coating system directionally like a tank coating system, involving epoxy or epoxy phenolics in at least a two coat application on an abrasive blast cleaned surface. Selection considerations include temperature resistance, abrasion resistance, and some service rating for immersion service. For warm insulation in particular, inorganic or organic zinc rich primers are avoided. Inspection of the surface preparation is critical in nonideal areas such as welds.

Visual inspection for the purpose of identifying the need to touch up failure points is not possible. Unless corrosion or insulation failure causes rework of the insulation entirely, there is no chance to do coatings work for 10 to 15 years, or more. Reluctance to spend resources on a coating system considering these limitations would be unwise, given the corrosion problems that often follow.

See Fig. 10.

Weather/Vaporproofing

The outer covering of the insulation system is a critical factor. First, it is the primary barrier to water that provides the corrosive environment. Second, it is the only part of the system that can be quickly inspected and economically repaired. The importance of desirable equipment design features was mentioned earlier. The following is a review of barrier properties as factors in CUI.

The purpose of a vapor barrier is to keep both liquid and vapor out of the insulation system. The purpose of a weather barrier, which should be used on warm equipment, is to keep liquid out, but permit evaporation of any liquid that gets in. For weatherproofing our standards require a minimum permeance of 115 $ng \cdot Pa^{-1} \cdot s^{-1} \cdot m^{-2}$ measured according to ASTM Test Method for Water Vapor Transmission Rate of Sheet Materials Using a Rapid Technique for Dynamic Measurement (E 398). Extensive use of metallic nonbreathing jacketing has probably contributed a great deal to our corrosion damage. Various types of mastic are applied to the breaks in jacketing systems, trying vainly to keep water out. With time, these seals are failing because of temperature limitations of mastics and aging characteristics. Liquid water entering at these breaks is evaporating in the insulation system with inadequate opportunity for vapor to escape.

FIG. 10—*Corrosion under fiberous insulation on another idle unit. In service, operating temperatures were high enough that paint was not considered necessary.*

At 100°C for example, each kilogram of water is going to produce almost 1.7 m^3 of vapor. Without a permeable weatherproof covering, the dew point in the insulation will equal the temperature of the hot face, and water will continually condense on the jacketing to be reabsorbed by the insulation. The small openings through which the water entered do not allow sufficient exit for the vapor.

There are other factors besides permeability to consider. These include durability and maintainability, appearance, contribution to fire protection (that is, melting point), flame spread resistance, and cost of installation. Like insulation selection, jacketing selection is a hard decision.

As mentioned earlier one of the reasons that the weatherproofing is such an important factor is that it is the only maintainable part of the system. Since mastics deteriorate quickly, a frequent schedule of inspection and maintenance is required. The weatherproofing or vaporproofing cannot be considered to last as long as the design basis of the whole system, say 10 to 25 years. In practice it must be maintained every 2 to 5 years to remain effective.

Climate

Both regional climate and microclimate should be considered factors of corrosion under insulation. Regional climate is important based on the reports of corrosion that tend to come from the more humid locations, especially where warm insulation is of concern.

Microclimate has to do with internal plant conditions, such as cooling tower drift and whether or not it is a salt water system, falling condensation from cold service equipment, subjection to steam discharges, spillage of process condensate, and so forth. It is often because of the microclimate to which an insulated item is subjected, that the worst corrosion and corrosion at elevated temperatures where it is normally not expected are found. Because of microclimate factors, the corrosion may be taking place continuously. The corrosivity of water that enters the insulation can also be increased, by cooling tower drift in particular.

Cooling tower drift is generally a very fine mist of water that can be carried airborne for 100 m or more, downwind. Cooling tower water is generally recirculated so that the original salinity is at least two to three times higher than the water supply. Except in locations very close to cooling towers, cooling tower drift water does not enter the insulation directly. Instead, it dries on the jacketing surface leaving a film of salts. Subsequent rain washes the collecting surfaces, carrying the concentrated salts to the equipment details, which are not weatherproofed effectively. There, they enter by gravity or by wicking action (in the case of absorbant insulation). As the cycles continue, salt concentrations continually increase as water is evaporated in the system.

Controlling microclimate, like equipment design, should be considered early, in this case, when considering equipment layout. In some cases it is possible to do something to eliminate the offending source after it is discovered. This is the preferred alternative. In some cases, weatherproofing can be upgraded to provide protection.

Maintenance Practices

With maintenance practices are included certain inspection practices for this discussion. As stated earlier, routine maintenance of weatherproofing is necessary to minimize defects in weatherproofing because of deterioration. Another critically important aspect is the making of major defects because of maintenance and inspection habits not oriented towards closure of the system promptly after work is completed. The use of contract insulation services, which are not always on hand when mechanical or inspection work is completed, probably contributes to this habit. In at least one case in our plant, openings made in the insulation of one of our vessels for ultrasonic inspection and inspection for external corrosion were never closed and are suspected as a major cause of later severe corrosion damage at the low points in that insulation system.

At our plant a very strong policy direction has been established that insulation openings will be closed promptly. All mechanical and inspection work cost estimates are to include insulation repair costs. This was not the case in the past, when the insulation work was estimated separately, and approved or not approved. If worked it would be several months and sometimes years later.

By comparison, in some of our units, loss or insulation would mean solidification of the process stream. There, a resident insulator follows mechanical work with immediate insulation repairs, and also pursues general insulation maintenance at other times. In this particular unit, temperatures are too high for corrosion as long as the weatherproofing is reasonably maintained. None

FIG. 11—*Mechanical work (a nozzle addition) has been completed. The insulation was left as shown for several weeks before insulation was closed. Note the deteriorating condition of the alkyd primer on the exposed steel surface.*

FIG. 12—*Inspection openings and mechanical damage that had been left unrepaired indefinitely until the present program started.*

the less, the visible contrast between this and other areas of the plant shows the benefit of a resident insulator in maintenance of insulation systems.

It should be noted that both the National Board Inspection Code and American Petroleum Institute (API) Pressure Vessel Inspection Code 510 require removal of some insulation at least every five years on all vessels where external corrosion is possible. In our plant alone we should be inspecting, very roughly 30 or 40 insulated vessels per year this way. The need for insulators in support of this inspection function should be obvious.

See Figs. 11 and 12.

Conclusion

In summary, there are numerous factors involved in causing or preventing corrosion under insulation. They have been grouped into seven categories and reviewed to show how they influence the three requirements for corrosion: exposure cycle, corrosivity of the water, and failure of coatings. Twelve illustrations were included, showing examples of equipment design details, maintenance and inspection openings, corrosion at entry points in hot insulation systems, and corrosion in moderate temperature insulation.

Dale McIntyre[1]

Factors Affecting the Stress Corrosion Cracking of Austenitic Stainless Steels Under Thermal Insulation

REFERENCE: McIntyre, D., **"Factors Affecting the Stress Corrosion Cracking of Austenitic Stainless Steels Under Thermal Insulation,"** *Corrosion of Metals Under Thermal Insulation, ASTM STP 880*, W. I. Pollock and J. M. Barnhart, Eds., American Society for Testing and Materials, Philadelphia, 1985, pp. 27-41.

ABSTRACT: Fundamental factors affecting the stress corrosion cracking (SCC) of stainless steels under thermal insulation will be reviewed. Specific topics are susceptible alloys, problem temperature ranges, sources of chloride ions, effect of halides other than chlorides, effect of geographical location, effect of potential, pH and buffering agents, mechanisms of concentration, and mechanisms of inhibition. Field experience with closed cell versus wicking type insulation will be discussed. The effectiveness of the weather barrier in preventing SCC under insulation will be discussed in light of maintenance procedure and detail design practice.

KEY WORDS: corrosion, austenitic stainless steels, insulation, stress corrosion

The practical urgencies of protecting plant equipment have forced corrosion engineers to base measures to prevent external stress corrosion cracking (ESCC) under insulation on a series of assumptions for which there is no solid data base. These assumptions will be reviewed, and their effect on the selection of preventive measures will be noted. A series of conclusions, some of them at odds with past practice, will be presented based on experimental studies conducted in other areas of corrosion science.

[1]Battelle-Houston Operations, 2223 West Loop South, Suite 320, Houston, TX 77027.

The Nature of Environment on Stainless Steels Under Insulation

Given that whatever environment exists beneath the insulation can cause stress corrosion cracking (SCC), and assuming for the moment that this cracking is fundamentally transgranular chloride SCC (an assumption that will be reconsidered later), the obvious deduction is that the environment operates in or cycles through the temperature ranges that allow crack propagation and provides a source of water, a source of chlorides, and a source of oxygen or other oxidizer. These conditions are necessary for any stress corrosion cracking to occur. The other two necessary conditions, a susceptible alloy and net tensile stress, come from fabrication and material selection processes.

An electrolyte must be present for cracking to proceed, and water is the most likely. Two sources of water are endemic to the external surfaces of process vessels: atmospheric moisture and city potable water, either as wash water or fire water. Which is more likely to find its way under the weather barrier over insulation? Rainwater is the more frequent but many engineers insist that dousing with wash water or fire water (during deluge system testing) are by no means unusual.

City potable water is typically near-neutral and oxygen-saturated with chloride contents varying widely at approximately 150 ppm. Mineral content will vary widely depending on the source.

Atmospheric moisture, which comes to us as rain, fog, mist, or dew, is perhaps less variable but more difficult to study than potable water. Considerable data have been generated by researchers concerned with atmospheric corrosion of steels.

It has been estimated that 1 L of rain falling from a height of 1 km washes 326 m^3 of air. As it falls, it absorbs atmospheric gases, becoming significantly more acidic and approaching a pH of about 5.5 in rural areas. Near industrial areas the pH of rainwater can approach 4.5. If heavy concentrations of sulfur dioxide are in the air, the pH may be between 3 and 4 in some local areas [*1*]. It will also, of course, become saturated in oxygen as it passes through the air.

As it falls, the rain also sweeps the air of suspended salts. The amount of sodium chloride suspended as airborne particles varies widely and depends strongly on the distance from the seashore. Measurements made in Africa and Russia indicate that rainwater within 1.609 km (1 mile) of the seashore may have 100-ppm chloride ion. The persistance of wind-borne salt particles is remarkable; rainwater has a fairly consistent 10-ppm chloride ion even several hundred miles inland (Fig. 1) [*2*]. If atmospheric moisture takes the form of mist or fog, the concentration of chlorides has been measured at 200 to 400 ppm within a mile of the coast.

So atmospheric moisture, which could serve as the source of electrolyte for SCC under insulation, will be an ambient temperature and pressure thin film of liquid with an oxygen content of 8 ppm, a pH of somewhere between 3 and 5.5, and a chloride content probably ranging from 10 to 100 ppm with excursions to 400 ppm.

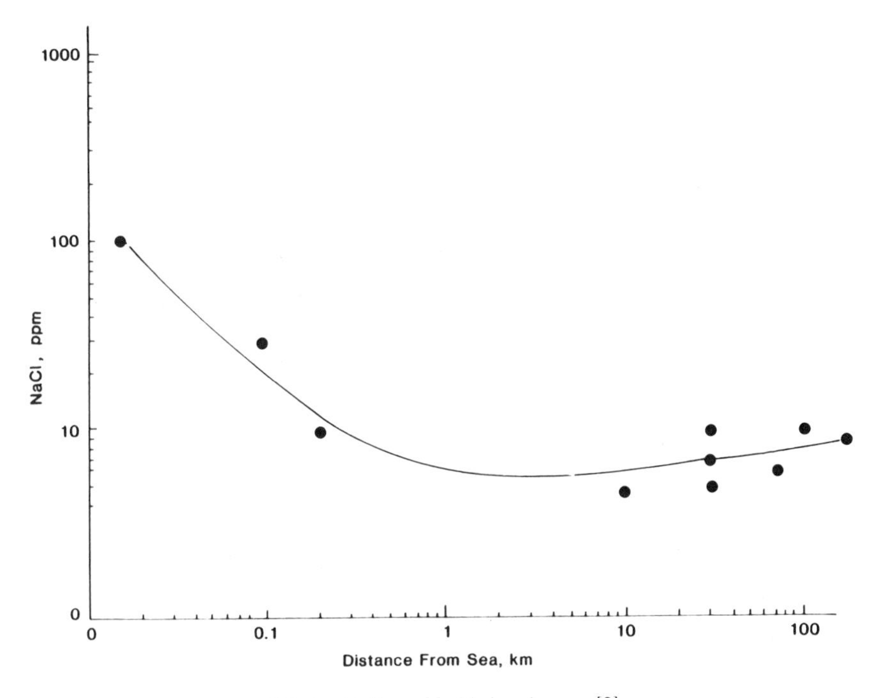

FIG. 1—*Sodium chloride in rainwater* [2].

Sources of Chlorides

When chloride stress corrosion cracking first surfaced in the closed pore insulation (CPI), it was assumed that the source of the chlorides was the insulation itself.

Is this valid? Certainly some of the old magnesia insulations were particularly high in chlorides, and certainly these materials would cause rapid failures of U-bends in the ASTM Evaluating the Influence of Wicking-Type Thermal Insulations on the Stress Corrosion Cracking Tendency of Austenitic Stainless Steel (C 692) test. Since then, most agencies and companies specify chloride contents in insulation that varies from 5 to 600 ppm [*3*]. How do these materials compare with atmospheric moisture as a potential source of chlorides?

If one imagines 0.0929 m^2 (1 ft^2) of a stainless steel vessel surface in the horizontal plane, under a nozzle, perhaps, on a column head, covered with 76.2 mm (3 in.) thick 2.24-kg/m^3 (14-lb/ft^3) calcium silicate insulation, the amount of chloride ion per unit area that could be leached out of that insulation is readily calculated as a function of its original concentration. Such data are shown in Fig. 2. If the insulation has 600-ppm bulk chlorides, the maximum allowed in the ASTM and Military (MIL) specifications, then 10 070 mg of chloride might deposit per square metre of stainless surface if all leachable chlorides

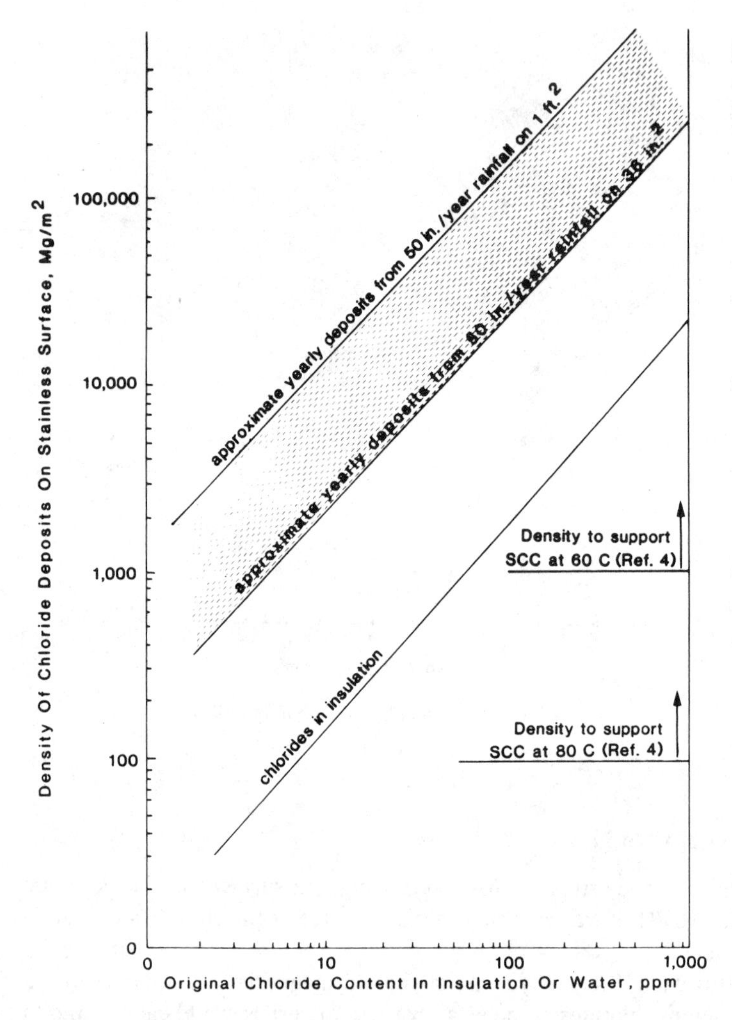

FIG. 2—*Potential density of chloride deposits from insulation or rainwater.*

ended up on the stainless. On the other end of the spectrum, insulation with less than 5-ppm chlorides in the bulk might deposit 84-mg Cl^-/m^2.

These numbers take on considerable significance when considered in the light of experiments done by Yajima and Arii of Japan's Toshiba Corporation [*4*]. At 80°C, these workers produced SCC in humidified air on stainless surfaces having deposited chloride levels of from 100 to 10 000 mg/m². This would suggest that even insulation with very low levels of chlorides (between 5 and 10 ppm) could still produce SCC if conditions of concentration were right.

How do these figures compare with chlorides transported by rainwater? Considering the hypothetical 0.0929 m² (1 ft²) of stainless surface, assume that its weather barrier has been completely ineffective, as they sometimes are around nozzles or after some aging. Rainfall in the Gulf Coast industrial regions ranges from 1016 to 1524 mm (40 to 60 in.) per year. Taking 1270 mm (50 in.) as an average, our 0.0929 m^2 (1 ft^2) of insulation would be exposed to 0.0928 m^2 $\times$ 1270 mm/year = 0.1179 m^3/0.0010 m^3/L = 118 L/year (144 $in.^2$ $\times$ 50 in./year = 7200 $in.^3$/61 $in.^3$/L = 118 L/year) of rainfall. On any facility more than about 1.609 km (1 mile) from the coast this rain would be expected to contain about 10-ppm chloride ion. The insulation itself cannot trap more than an absolute maximum of its own volume of this water, which for a 76.2 mm (3 in.) thick block 0.0929 m^2 (1 ft^2) in area would be about 7 L. Thus 111 L of rainwater must somehow find its way through the insulation block per year, and these 111 L will carry 111 000 $\times$ 10 $\times$ 10^{-6} = 1.1 g of chloride. If this salt is all deposited on the underlying stainless steel this would result in 11 721 mg of Cl^- per m^2 per year. On facilities closer to the coast, for example, marine terminals, up to 117 200 mg/m^2/year might be expected.

A 0.0929-m^2 (1-ft^2) rip in the weather barrier may seem unrealistic. Consider, then, a 38.7-cm^2 (6-$in.^2$) defect. Falling rain must first saturate the entire insulation block, not just the exposed area, before moisture can begin to drip onto the stainless. Even so, the calculated chloride density is ten times greater than that from insulation of equal chloride content.

Evaporation of rainwater results in some transfer of deposited chloride back to the atmosphere. Actual measured rates of chloride deposition on exposed surfaces range from 10 to 1000 mg/m^2/day [*2*].

These values suggest that chlorides leached from insulation can indeed cause external SCC if conditions of concentration are right. However, chlorides deposited from the atmosphere are potentially many times more dangerous. The density of atmospheric chlorides is higher and, unlike the insulation itself, the potential supply is infinite.

Many engineers argue that, as a source of chlorides, wash water or water from testing deluge systems cannot be ignored. Such exposure is difficult to quantify, since the amount of water impinging on the exposed surface of the insulation will vary widely depending on the interest and enthusiasm of the worker holding the wash hose. However, again taking 0.0929 m^2 (1 ft^2) of exposed calcium silicate insulation, if 114 L (30 gal) of potable city water are allowed to run into the surface per year, and assuming a chloride content of 150 ppm, the potential maximum density of deposited chlorides is 171 000 mg Cl^- per m^2 per year, worse even than seacoast rainwater. However, the actual amount of chlorides deposited would be significantly less than the calculated value since the relatively brief, intense duration of washing, or deluge-system testing, would cause more runoff and less evaporation and concentration than the more gradual accumulation of moisture from rainfall. To counter this, the more intense runoff would also result in greater dilution and depletion of any inhibitor deposited on the surface.

Polyvinyl chloride (PVC) plants and some other chloride-based processes are of themselves significant sources of chlorides. Resin fines can often be seen covering vessels and piping to a depth of several millimetres in some PVC plants; breakdown of this PVC because of water and heat can cause ESCC in plants sited hundreds of miles from any seacoast.

Sources of Water

The work of Yajima and Arii [*4*] also brings focus on the second fundamental assumption that corrosion engineers must make regarding stress corrosion cracking under insulation. What is the mechanism by which water comes in contact with the stainless surface? Is it rainwater soaking through the insulation at defects in the weather barrier? Or does condensation or droplet formation during periods of high humidity play a part? How about runoff from other areas of the vessel?

The assumed answer to this question (and there are only assumed answers) is affected by assumptions to the first question, and in turn affects the choice of preventive measure. If rainwater only is to be feared, and that rainwater must soak through the insulation to contact the surface, then the insulation system's efficiency as a weather barrier determines its ability to prevent SCC, and the use of inhibited insulation is logical.

If, on the other hand, humidified air alone is sufficient to cause SCC, then to prevent SCC the insulation system must not only function as a weather barrier but also as a vapor barrier to prevent moisture-laden high-humidity air from contacting the stainless surface. There was a time, when thick tarry asphaultic insulation covers were used, when an effective vapor seal may have been achieved. Today's metal foil weather sheeting, however, was never designed as a vapor barrier but only as a weather barrier. Thus, if an effective vapor barrier is necessary, some sort of barrier coating or sacrificial coating will be required.

Yajima and Arii covered thin stainless steel tubes with salt densities varying from 100 to 10 000 mg/m^2. They then exposed these tubes in humidity cabinets where temperatures could be controlled at 50 to 80°C and relative humidities from 60 to 80%. They produced SCC at 80°C at all salt densities and relative humidities down to 60%. At high salt densities they observed SCC at 60°C down to relative humidities of 70%.

These data suggest that in the presence of deposited salt humid air by itself might be sufficient to cause SCC. How can this be?

Two mechanisms suggest themselves. Droplet formation is possible even at relative humidities below saturation if the surface temperature is lower than some value related to the absolute water content.

Figure 3 presents a series of curves showing the temperature difference that would cause droplet formation at various relative humidities and ambient temperatures. This effect may be significant on many vessels in cyclic service,

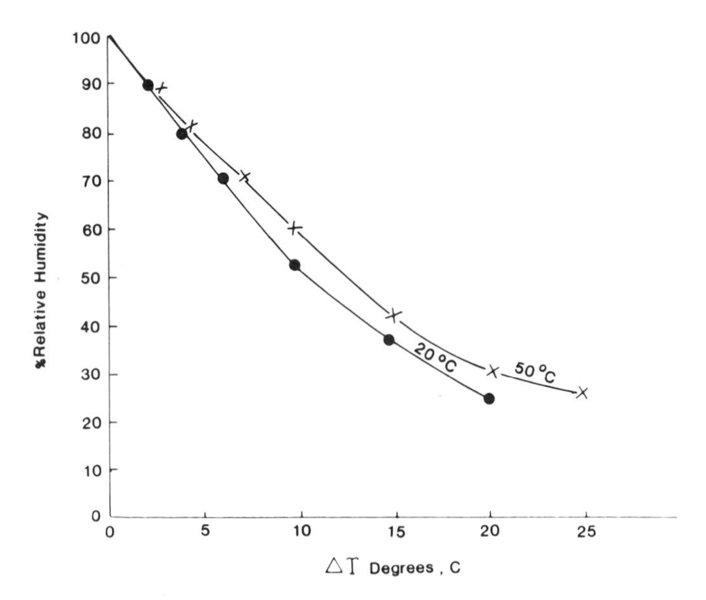

FIG. 3—*Temperature difference* ΔT *between stainless surface and ambient air temperature for droplet condensation* [1].

whereby droplets could condense when the pipe or vessel is cool, then evaporate when the vessel is heated.

Another mechanism revolves around the hygroscopic nature of salt deposits. Figure 4 presents data showing the weight gain of sodium chloride at ambient temperature and various relative humidities as a function of time. Note that after two weeks at high relative humidities the salt had absorbed over three times its original weight of water. Such a deposit should cease to be a salt particle and should become in essence a droplet of saturated brine, with an expected conductivity on the order of 210 $S \cdot cm^{-1}$, a better electrolyte than seawater.

Data from Fig. 4, combined with the results of Yajima and Arii, suggest that stainless steel with salt deposits (resulting perhaps from simple atmospheric exposure after erection and before insulation is applied) might suffer SCC even under an intact weather barrier if high-humidity air could reach the hot stainless surface.

Neither of these mechanisms has been proven to general satisfaction. Certainly the majority of SCC failures under insulation occur in areas where the weather barrier has broken down, and where the insulation is actually wet to the touch. However, such mechanisms are useful for understanding failures observed on vessels inside buildings, which are never exposed to rainfall, and the preservice cracking of vessels shipped by sea to high-temperature areas such as the Middle East.

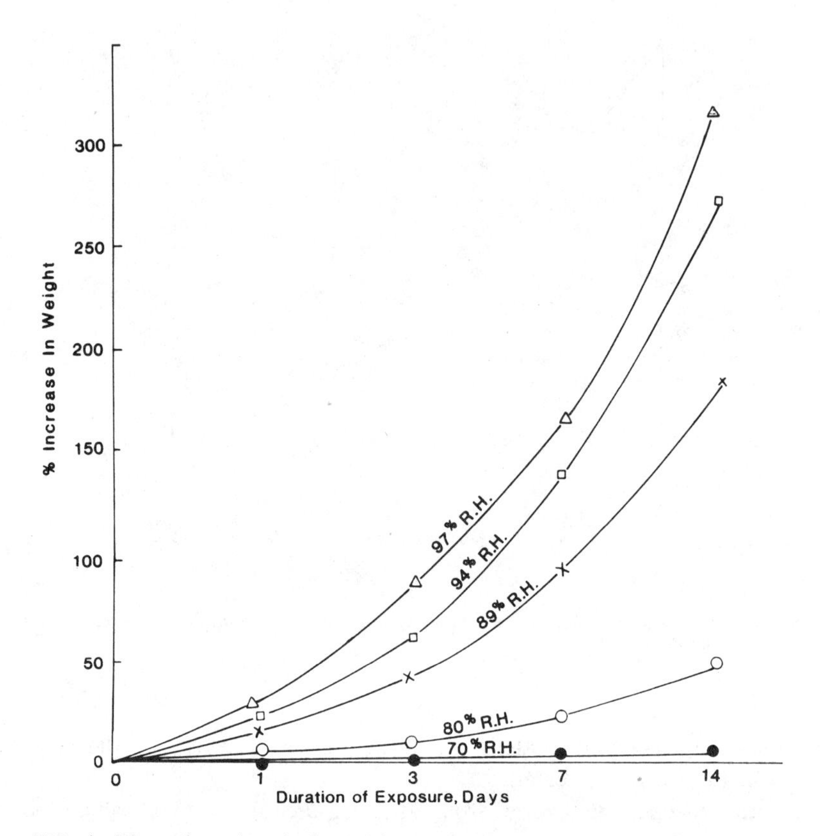

FIG. 4—*Water absorption of salt particles as a function of time and relative humidity* [2].

The Role of Potential and Inhibitors

Potential is now known to be of fundamental importance to SCC. The oxygen-saturated rainfall or condensation to be expected on stainless surfaces under insulation will have an open circuit potential squarely in the range of cracking. In both dilute and concentrated chloride solutions the critical potential for cracking appears to be less negative than about $-0.200\ V_{SHE}$ (where SHE is standard hydrogen electrode) thus cracking can be avoided by forcing the potential more negative than $-0.200\ V_{SHE}$. This can be done either by lowering the oxygen content of the solution below about 100 ppb or by driving the potential more negative with cathodic protection or inhibitor additions.

Lowering the oxygen content in the atmospheric environment would not seem possible, but that is precisely the benefit provided by some very effective barrier coatings. All organic coatings are permeable to some extent; however, the nature of the polymers gives great variation in permeability to different species. For example, the widely used epoxy coatings are fairly permeable to water. However, they have an extremely low permeability to oxygen and thus are an effective preventive measure in dilute atmospheres. (Note: epoxies are sometimes less effective in, for example, PVC plants, where the breakdown of PVC dust forms hydrochloric acid, which does not need oxygen to cause cracking.)

Some inhibitors and all cathodic protection systems use the potential shift as their primary preventive measure. Their effectiveness on an experimental level is well documented; the life of the systems in the field depends on several details of application.

For example, sodium metasilicate inhibited insulation is widely used both in the United States and in Europe. The inhibitor concentration is often specified as some ratio of the chloride content. Figure 5 reproduces a widely used graph from Ref 5, data that have been written into ASTM and MIL-I-24244 specifications. The ratios of inhibitor to chloride are typically at least 10 to 1.

These data were developed on sensitized test specimens, with the acceptance criterion that not more than one out of four of the samples would be cracked after 28 days. How many plant managers would accept a failure probability of 0.25?

The basic assumptions when using inhibited insulation are that the water to cause cracking will seep through the insulation, leaching out inhibitor as well as chloride, and that any concentration mechanism that subsequently concentrates the chlorides will also concentrate the inhibitor. The first assumption is

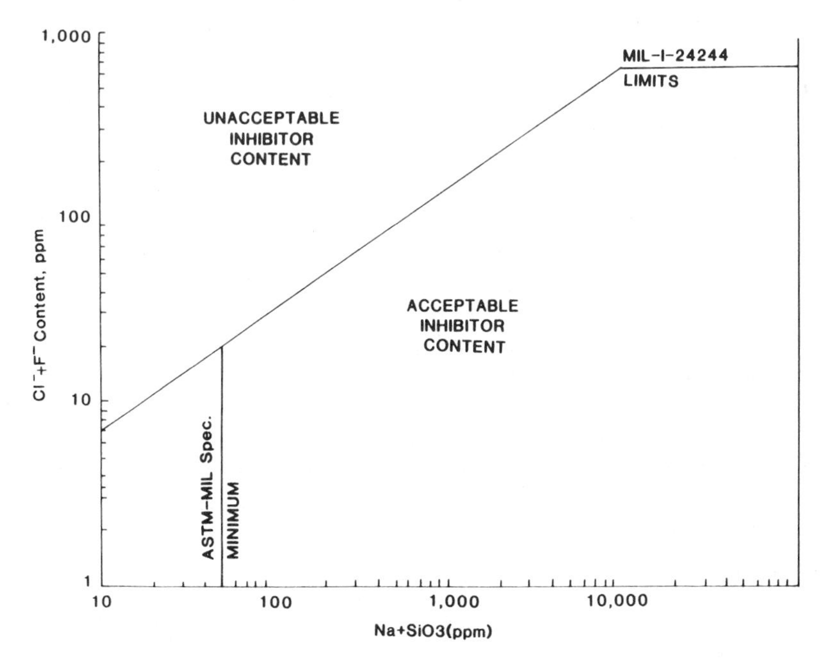

FIG. 5—*Inhibitor to halides ratios required by MIL specifications and the U.S. Nuclear Regulatory Agency* [5].

certainly not true for vessels in cyclic service, where condensation of droplets directly on the stainless surface may play a role in cracking. Runoff water, whether rainwater or wash water, can find its way to the stainless surface at breaks in the caulking without actually soaking through the insulation. If, as has been suggested, droplet formation at hygroscopic salts can take place, once again the water, not having passed through the insulation, will not be inhibited.

The assumption that inhibitor will concentrate on the hot stainless surface is worth examining in the light of the extremely high solubility of sodium metasilicate in water. Consider again the hypothetical 0.0929 m^2 (1 ft^2) of exposed insulation covering 0.0929 m^2 (1 ft^2) of stainless steel. If rainwater is allowed to impinge on this surface, what will happen? After approximately 7 L of rain have fallen, the insulation will be more than saturated with water. After that, some of the rain will run off the surface, some will reevaporate back out of the insulation, and certainly some will make its way through the insulation, carrying with it chlorides and inhibitor.

Inhibitor concentrations in insulation that will pass the ASTM C 692 test vary from 1.5 to up to 20% by weight of the insulation. In 0.0929 m^2 (1 ft^2) of 76.2 mm (3 in.) thick insulation, there would thus be from 24 to 318 g of inhibitor. The initial 7 L of rainwater it would take to saturate 0.0929 m^2 (1 ft^2) of insulation, would therefore have an inhibitor concentration of from 0.34 to 4.5%. The solubility of sodium metasilicate in water exceeds 5 equivalents per litre, or 30%. Therefore, would not essentially all the inhibitor go into solution in that first 7 L of rainwater? Why would there be any sodium metasilicate left to inhibit the remaining 111 L of rainwater expected to fall on the hypothetical square foot of insulation in the course of a year?

Assuming that the inhibitor leached from the insulation is reconcentrated on the surface, very high densities of inhibitor will contact the stainless steel, from 253 000 to 3.3 million mg/m^2. However, this is not renewable. Chloride deposition from rainwater near the coast, or from the assumed 0.113 m^3 (30 gal) of wash water or deluge-test water spilled on the vessel per year, accumulating at rates as shown in Fig. 2, would result in unsafe inhibitor to silicate ratios of less than 10 to 1 in about three months with 15 000 original ppm sodium silicate in the insulation. For the highly inhibited 20% sodium silicate material, the inhibitor to chloride ratio could drop below 10 after something less than three years. For the more typical case of plants sited more than about 16-km (10 miles) inland, inhibitor depletion would obviously take longer if rainwater is the source of the chlorides, from about 2.5 years on insulation with low levels of inhibitor up to a calculated 25 years at high inhibitor concentrations.

However, this assessment ignores the possibility of the inhibitor being dissolved off the stainless surface after deposition. Evaporation at 60 to 80°C, although rapid compared to room temperature, is still slow enough to allow some runoff if enough water seeps through the insulation. Once again, the high solubility of sodium metasilicate in water makes it possible to lose significant quantities of inhibitor.

Therefore, the use of inhibited insulation involves not only adequate initial concentrations of inhibitor but also careful monitoring of these concentrations as a function of time, with replacement of insulation sections that become depleted. This involves significant maintenance effort for successful long-term use.

Other preventive measures, barrier coatings and sacrificial coatings or foil wraps, are not limited by the assumption that water must seep through the insulation, or the assumption that the insulation is the predominant source of chlorides. Barrier coatings and sacrificial methods are designed to protect, regardless of the source of the water or the source of the chlorides.

The success of barrier coatings will clearly depend on defect density. Companies who have tried barrier coatings have quoted widely varying failure rates, from less than 1 to 75%. Other papers in this publication address this issue.

The success of sacrificial methods, either metal-loaded coatings, thermal spray, or foil wrap, will depend on the achievement of continuity between the sacrificial anode and the stainless cathode through what is undoubtedly an intermittent, discontinuous electrolyte. Once again other papers in this session will address that issue.

Temperature Ranges for SCC

The assumption of a temperature range in which cracking occurs must be made to decide which items must be protected. There is, unfortunately, uncertainty on both the low and the high end of the temperature range. Some engineers consider equipment that operates above about 121°C (250°F) to be safe since water flashes off exposed surfaces very quickly at such temperatures. However, a number of failures have been recorded on vessels operating at or around 121°C (250°F). Dillon [*6*] states that one must go higher, to at least 260°C (580°F), where chloride salts lose their water of hydration, before one can reasonably expect stainless steel to be free of stress corrosion under steady state conditions. One European firm specifies preventive measures on all equipment operating below 538°C (1000°F). If equipment operates above the assumed limit for SCC, the effect of temperature cycling during start-up and shutdown must still be considered.

The lower temperature limit for stress corrosion cracking is more difficult to establish than the upper limit. The crack growth rate of stainless steel in chlorides increases radically above 80°C. But below 80°C, the observed threshold temperature appears to depend on the patience of the observer. Some engineers use 70°C. Others use 65, 60, or 55°C, depending on their experience and their willingness to accept risk.

The search for a lower temperature threshold must ignore many cracks documented in vessels operating at ambient temperatures or below. Many of these stress corrosion cracks have formed after 20 or 25 years of service, but they have formed on vessels or piping that were never heated above Texas ambient. Dillon [*7*] refers to stress corrosion cracks that occurred at cryogenic temperatures in organic chloride service. It seems clear that there is nothing sacred about 60 or 55°C that prevents stress corrosion cracking. Rather the lowering of temperature gradually increases initiation time and reduces the crack propagation rate of SCC until other operational problems, such as pitting, control the life.

Along these lines the data of Yajima and Arii [*4*] are pertinent. Their experiments on thin-walled pressurized American Iron and Steel Institute (AISI) 304 (Unified Numbering System [UNS] S30400) stainless steel tubes were conducted at varying temperatures in a humidity cabinet with sodium chloride on their surfaces. Cracking occurred rapidly above 70°C with no apparent pitting. At 60°C, failure was still by SCC although some pitting was evident. At 50°C, with some iron chloride on the surface, pitting predominated although some SCC was observed on microscopic examination. Thus we see again a continuum rather than a threshold phenomenon.

Susceptible Materials

It has been clear for 40 years that the austenitic stainless steels have a particular susceptibility to chloride stress corrosion cracking. The vast majority of problems have been observed on the 300 series stainless steels, simply because these alloys make up about 50% of the world's stainless steel production. Failures have been recorded on equipment of AISI Types 304, 304L, 321, 347, 316, and 316L (UNS S30400, S30403, S31600, S31603) stainless steels. Are there any significant differences in susceptibility between these types? Intuitively one would expect that the nonmolybdenum bearing grades, with their known higher susceptibility to the pitting that so often provides the stress raisers from which SCC propagates, would be more susceptible than the more highly alloyed grades. Would the nonmolybdenum bearing grades therefore justify more extensive protection? Laboratory evidence on this point is contradictory. Some data collected by operators of PVC plants would suggest that Type 316 is indeed somewhat more resistant [*7*]. For the more general diversified plant no convincing statistics have been advanced to support any significant difference in susceptibility.

Are the L-grade materials more resistant than regular-carbon materials? No laboratory researcher has ever been able to show that low-carbon materials resist transgranular SCC better than regular-carbon materials. However, reports of external intergranular SCC are becoming more and more common on regular-carbon vessels and piping even in ambient temperature applications, or during shipping and storage. How many SCC failures of regular-carbon stainless steel near welds have been replaced without analysis on the assumption that it was transgranular cracking and not to the sensitization known to occur during welding?

One effect that has not been adequately researched is the effect of weld metal composition on SCC resistance of nearby base metal. Weld rods for austenitic stainless steels are almost always balanced to give some ferrite on cooling by silicon additions. Silicon is known to shift the potential of alloys to which it is added in the noble direction. Such a shift will not hurt the resistance of the welds themselves, which are known to be slightly more resistant than their base metal to SCC. But what about the galvanic effect on the adjacent base metal? Surely the effect will be small, no more than 50 or 100 mV. But 100 mV of anodic potential shift has been shown to profoundly affect the SCC of many alloys in many different environments.

Structural Integrity of Cracked Vessels

One last assumption needs to be examined; the assumption that if external stress corrosion cracking occurs it will take the form of leakage or pinhole weeping. The great preponderance of field experience suggests that this is true; the excellent toughness of the 300 series stainless steels is very forgiving. However, at least three cases of gross loss of strength have been documented. In one case from a PVC plant, a 304 stainless steel heat exchanger flange became so weakened by external SCC that pieces could be broken off by hand. In another case, a 203 mm (8-in.), schedule ten, 304L stainless steel pipe suffered a catastrophic rupture approximately 0.6096 m (2 ft) in length upon hydrostatic test at 1034 kPa (150 psi) [8]. In a third case, a 0.4572 m (1.5 ft) long piece of stainless steel simply fell out of a 304L stainless steel resin slurry tank in a PVC plant [7].

None of the events described above caused serious losses. However in lethal service any of these three incidents could have caused a major release. They serve as a warning that the structural integrity of stainless steel vessels or piping with extensive stress corrosion cracks cannot be taken for granted.

Discussion

The considerations set forth above suggest a modification or supplementation to traditional methods for testing insulation regarding ESCC. In particular, the SCC preventive system must be addressed rather than just the insulation. One suitable test, first proposed by Ashbaugh in 1967, would retain the use of heated U-bends but would introduce test solution to seep through the insulation from the top rather than wick up through the bottom. The test solution would contain 100-ppm chloride as sodium chloride and be acidified to a pH of 4.5. The ratio of the total test solution volume to the volume of the insulation block should be at least 10 to 1; the report of test results should include a phenolthalein spot test for sodium metasilicate inhibitor on both the stainless surface and in the insulation itself. Test solution drip rate and total test time could be determined by round robin testing or a project undertaken by a group industry association such as MTI or ASTM.

Some of the engineering concerns regarding stress corrosion cracking under insulation have never been satisfactorily resolved. Major operational questions remain, including:

1. How can a major vessel or piping run be inspected economically for external stress corrosion cracking without completely stripping the insulation during a total shutdown?
2. If stress corrosion cracks are located by dye-penetrant, how can they be characterized as to depth?
3. How can the structural integrity of a vessel with extensive stress corrosion cracking be analyzed to either assure against catastrophic failure, or justify replacement?
4. How can existing stress corrosion cracks be arrested until a production window allows repair or replacement?

These are vital questions for plants trying to live with aging vessels and piping runs. Approaches have been formulated to address each of these concerns; perhaps they might be worthy of a symposium such as this in future years.

Conclusions

Study of available data, which are by no means complete, suggest the following conclusions:

1. Even insulation with only 10-ppm chloride might cause chloride densities high enough to support stress corrosion cracking if the chlorides are leached out and deposited on the stainless surface.
2. However, the density of potential deposits of environmental chlorides (from rainwater, droplet condensation, or wash water) is 10 to 100 times greater than leachable chlorides in the insulation itself. Environmental chlorides would therefore appear to be more of a threat than chlorides in insulation.
3. Although the most common mode of failure is leakage or seeping, catastrophic failure of stainless steel vessels stress-cracked under insulation is possible and has been documented.
4. Although the most common source of the electrolyte to support cracking appears to be water ingress (of rain, wash water, or fire water) at defects in the weather barrier, stress corrosion cracking under intact weather barriers is possible on vessels in cyclic temperature service or on surfaces previously contaminated with hygroscopic salts.
5. A successful long-term preventive measure for stress corrosion cracking must be able to handle chlorides both from the insulation and from other external sources, and must be able to cope with water contacting the stainless

surface after either seeping through the insulation, condensing on the surface as droplets, or running along the surface from remote access points.

References

[1] Rocenfeld, I. L., *Atmospheric Corrosion of Metals*, National Association of Corrosion Engineers, 1972, p. 97, 203.
[2] Rocenfeld, I. L., *Atmospheric Corrosion of Metals*, National Association of Corrosion Engineers, 1972, p. 98.
[3] Nickolson, J. D., "Application of Thermal Insulation to Stainless Steel Surfaces," *Bulletin of the Institution of Corrosion Science and Technology*, Vol. 19, No. 5, Oct. 1981.
[4] Yajima, M. and Arii, M., "Chloride Stress Corrosion Cracking of AISI 304 Stainless Steel in Air," *Materials Performance*, Dec. 1980, pp. 17-19.
[5] Gillett, J. and Johnson, K. A., "An Experimental Investigation inot Stress Corrosion Cracking of Austenitic Stainless Steel Under Insulation," *Corrosion Under Lagging*, Institute of Corrosion Science and Technology, Newcastle-upon-Tyne, United Kingdom, Nov. 1980.
[6] Dillon, C. P., *Guidelines for the Prevention of Stress-Corrosion Cracking of Nickel-Bearing Stainless Steels and Nickel Alloys*, MTI Manual 1, Materials Technology Institute, 1979.
[7] "Report on Vinyl Chloride Safety Association Meeting," Sept. 1983, Toronto, Canada.
[8] Dillion, C. P., Ed., *Forms of Corrosion: Recognition and Prevention*, National Association of Corrosion Engineers, 1982, p. 66.

James A. Richardson[1]

A Review of the European Meeting on Corrosion Under Lagging Held in England, November 1980

REFERENCE: Richardson, J. A., **"A Review of the European Meeting on Corrosion Under Lagging Held in England, November 1980,"** *Corrosion of Metals Under Thermal Insulation, ASTM STP 880,* W. I. Pollock and J. M. Barnhart, Eds., American Society for Testing and Materials, Philadelphia, 1985, pp. 42–59.

ABSTRACT: The Meeting on Corrosion Under Lagging was sponsored by the Institution of Corrosion Science and Technology, the Thermal Insulation Manufacturers' and Suppliers' Association, and the Thermal Insulation Users' Liaison Group. Ten papers were read representing the interests of users, suppliers/contractors, consultants, and inspection engineers, and the abstracts of these papers are recorded in the appendix of this paper. Additionally, the themes, lessons, or perspectives arising out of the meeting are reviewed in the areas of corrosion phenomena/case histories; corrosivity of wet insulation/fireproofing; corrosion prevention using inhibited laggings, organic coatings, metallic foils, and paints and alloy fabrication/selection; and corrosion prevention through design, specification, inspection, and maintenance. Areas where a consensus emerged are highlighted.

KEY WORDS: carbon steels, low alloy steels, austenitic stainless steels, calcium silicates, insulation, cellular glass, ceramic fibers, mineral wool, corrosion, stress corrosion, organic coatings, aluminum foil, stainless steel foil, specifications

The original impetus for the Meeting on Corrosion Lagging came from the Thermal Insulation Users Liaison Group in the United Kingdom. They were concerned that a communication gap had developed between the thermal insulation supplier/contractor and the user on those aspects of specifications relating to corrosion control. Suppliers and contractors were not in all cases familiar with the various corrosion mechanisms that can operate under lagging systems, and were therefore not necessarily "sympathetic" towards corrosion control specifications "thrust upon them" by the user, with consequent

[1] Senior corrosion engineer, Engineering Department, North East Group, Imperial Chemical Industries PLC, P.O. Box 6, Billingham, Cleveland TS23 1LD, United Kingdom.

effects on achieved standards of application. On the other hand, user corrosion engineers were too inclined to view corrosion in terms of one aspect of the lagging system rather than the system as a whole and had thus contrived to produce an almost bewildering array of corrosion control specifications, which in some cases entertained an unrealistic concept of what was achievable under site application conditions. It followed that the meeting's objectives were as much mutual education or information as breaking new ground in the technical sense. It was hoped that it would provide a state of the art review of current European practice and stimulate productive dialogue between the various interested parties.

Given the objectives of the meeting, it seemed appropriate that it should be organized by the two bodies representing corrosion engineering and the insulation supply/contracting industry in the United Kingdom, and the task was taken on jointly by the Institution of Corrosion Science and Technology and the Thermal Insulation Manufacturers' and Suppliers' Association.

The meeting was a considerable success. Most European countries were represented in the more than 100 delegates who attended. Ten papers were read representing the interests of users (4 papers), suppliers/contractors (3 papers), consultants (2 papers), and inspection agencies (1 paper). The abstracts of the papers, where available, are provided in the Appendix of this paper.

This paper attempts to review the two-day meeting, and in particular any themes, lessons, or perspectives arising out of discussion. It is inevitably "one man's view" of a wide ranging meeting and may well reflect some of the biases/preferences of that man, albeit the Chairman of the Organizing Committee. Rather than review papers individually, it is considered more appropriate to review the meeting in terms of a number of specific technical topics that arose when discussing corrosion under fireproofing and thermal insulation systems. Individual papers are referred to by their number in the enclosed Appendix. It is inappropriate in a review such as this to draw conclusions on specific technical issues. However, those areas where a large measure of consensus prevailed at the meeting are indicated.

Corrosion Phenomena and Case Histories

In the event, the meeting restricted itself to consideration of carbon, low alloy, and stainless steel substrates. Accepting that water can penetrate lagging or fireproofing systems, Paper 2 anticipated the following possible problems, and described their phenomenology/mechanisms in basic terms.

1. *Carbon/Low Alloy Steels*—The most significant problem is likely to be pitting or general corrosion. These materials are normally "passive" in alkaline environments, and might thus not be expected to corrode in, for example, water extracts from cementitious fireproofing, or calcium silicate/magnesia

based thermal insulants. However, specific anions, notably the chloride ion, which can arise either from the materials themselves, or as airborne or waterborne contaminants, are able to break down passivity locally and initiate pitting corrosion. If penetration by acidic airborne or waterborne contaminants, for example, sulfur or nitrogen oxides, is possible, or if water extracts from the lagging materials are acidic as in the case of certain organic cellular foams, then general corrosion is likely to ensue. Certain specific air or waterborne contaminants, notably the nitrate anion, can give rise to external stress corrosion cracking of nonstress-relieved systems, particularly if a cyclic wetting or drying concentration mechanism prevails. The consensus was that a plant operating continuously or intermittently within the temperature range approximately −5 to 200°C is at risk, although opinions varied on the limits (particularly the upper limit) and on the definition of "intermittently."

2. *Stainless Steels*—By far the most significant problem is external stress corrosion cracking of austenitic stainless steels. This is caused by chloride ions that originate either in the insulation or fireproofing materials themselves, or as airborne or waterborne contaminants. The simple expedient of thermal stress relief, normally so successful in preventing stress corrosion cracking (SCC) of carbon or low alloy steel, systems, is not normally practicable in austenitic stainless steel systems. Surprisingly small levels of chloride are sufficient to cause the problem, particularly if a cyclic wetting or drying concentration mechanism prevails.

The consensus was that a plant operating continuously or intermittently above approximately 60°C is at risk. There was some divergence on the significance of the risk at temperatures at approximately >200°C, on the grounds that surfaces normally hotter than this can spend but a fraction of their life wet. However, there was a consensus that most significant problems had occurred on surfaces operating normally at temperatures <200°C.

Case histories of specific carbon steel corrosion problems were presented in Paper 8, and other cases were cited in the discussion. Corrosion beneath cementitious fireproofing had been experienced at locations in the Caribbean Sea, United Kingdom, United States, Singapore, Japan, and Saudi Arabia. Corrosion had usually, but not always, been exacerbated by the presence of sea salts caused by either their inclusion in the originally cast concrete or by their progressive accumulation from the atmosphere at coastal sites. In one case, 50% metal loss had been experienced in 16-years service. Corrosion under thermal insulation had been experienced under a range of lagging materials at different temperatures. In one case, significant corrosion had been experienced after 12 years under a cold lagging system (−5°C) on some tanks, particularly on the roof and the first two shell courses. The insulant was in-situ foamed polyurethane under a galvanized steel vapor jacket. In another case, corrosion of a carbon steel heavy fuel oil pipeline operating at approximately 120°C beneath preformed calcium silicate insulation, metal clad, had resulted in hydrocarbon leakage and a significant fire.

Case histories of external stress corrosion cracking of austenitic stainless steels were presented in Paper 3, and again others were cited in the discussion. The major case study concerned failure of brewery vessels and piping systems operating at $<100°C$. Major vessel failures had resulted from flooding (because of inadequate weatherproofing) of polyurethane foam lagging containing chlorinated organic fire retardants, hydrolysis of which yielded significant quantities of free chloride. Other failures were reported under calcium silicate, and under virtually chloride-free insulants including fiberglass blanket and asbestos. The majority of failures reported were from the brewing or dairy industries, probably because of the large population of welded stainless steel equipment from these industries operating in the most critical temperature range. In this context, an interesting paper from New Zealand [*1*] was submitted for discussion at the meeting (Fig. 1). It reported on 77 cases of stainless steel corrosion from all sectors of New Zealand industry. Figure 1 is taken from the paper. Many of the "environmental cracking" cases reported were, in fact, cases of external stress corrosion cracking under lagging, and their impact in terms of proportion of total failures and failure costs is self evident.

Corrosivity of Wet Insulation or Fireproofing

Two issues were discussed at the meeting:

(1) the composition of the wet lagging/fireproofing environment in relation to its corrosivity and

(2) the physical transport of aqueous fluids through lagging or fireproofing systems.

Fireproofing was discussed in Paper 8. Cementitious coatings furnish an alkaline environment, pH 11 to 12, which would normally be passivating towards steel. Soluble chloride is the usual promoter of corrosion, and it arises either from the use of calcium chloride accelerator or antifreeze or salt water in the preparation of the mix, or as an airborne or waterborne contaminant. Chloride levels as low as 700 ppm may prove sufficient to initiate corrosion. Penetration of concrete occurs either through the matrix, where an excessive water/cement (W/C) ratio has been used, through cracks associated with shrinkage, thermal fluctuations, or mechanical damage, or at junctions with steel equipment.

The wet lagging environment was addressed specifically in Papers 1, 3, 4, 7, and 8, and arose in discussion. A variety of analytical data were presented from leaching tests on lagging materials undertaken at varying water/solid ratios and temperatures. Unsurprisingly, there was considerable scatter in the data, but the following consensus was identifiable:

1. Calcium silicate, fiberglass, cellular glass, and ceramic fiber furnish neutral to alkaline environments when wetted, with pH values in the range 7

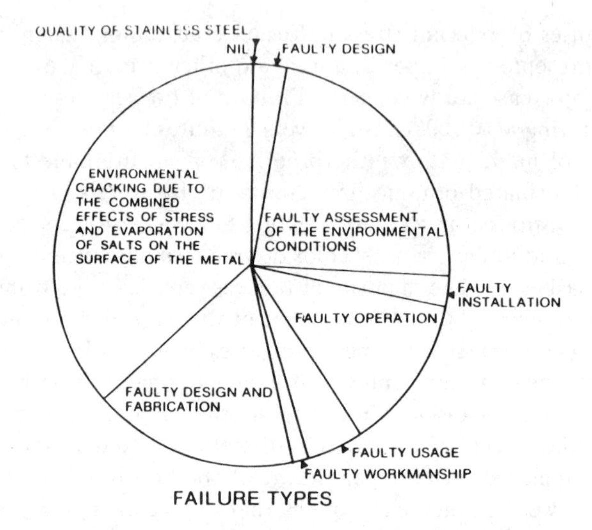

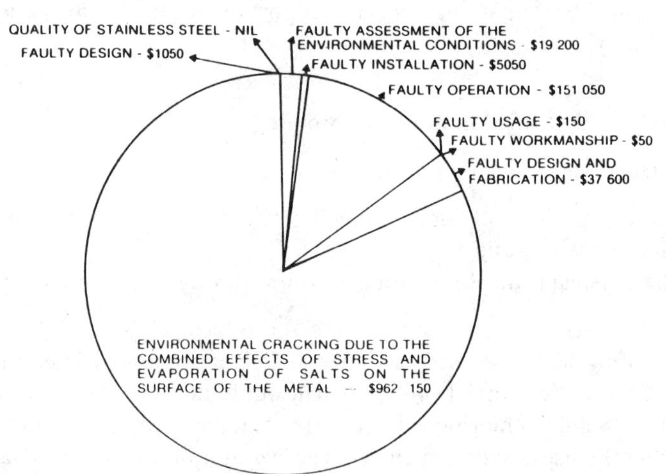

FIG. 1—*Categories under which stainless steel failures can be grouped and the costs of corrosion within these categories* [1].

to 11. Cellular glass is free of soluble chloride. The other materials can contain chloride. For example in the United Kingdom, The British Standard for Thermal Insulating Materials (BS 3958, Part 1), for preformed calcium silicate allows "up to approximately 500-ppm chloride, and a figure of 20-ppm chloride on a water extract from calcium silicate" was quoted in discussion.

2. Mineral wool furnishes essentially neutral environments when wet, typically pH 6 to 7. Chloride levels are low, one figure of 2 ppm on a water extract being quoted.

3. Water extracts from organic foams can be appreciably acidic, pH values as low as 2 to 3 being readily achievable. Additionally, where halogenated fire retardant species (typically halogenated phosphate esters) have been included in the foam, water extracts can show very high levels of free halide, up to thousands of ppm, dependent upon the degree of hydrolysis achieved under the test conditions, principally, time and temperature.

The data presented in Paper 8 were as representative as any, and Tables 1 through 3 are reproduced directly from the paper. Carbon steel corrosion rates as high as 15 to 20 mpy (where mpy is mil per year) were reported for organic foam aqueous extracts, and some of the more significant stainless steel external SCC problems reported at the meeting were associated with fire-retarded organic foam insulations.

TABLE 1—*Properties of room temperature aqueous extracts from polyurethane and phenolic foams* [3].

Properties	Phenolic Foam	Polyurethane Foam
pH free water	3.45	6.1
pH water in foam	...[a]	...[a]
Water pickup, g	13	0.1

[a] Insufficient water absorbed by the foam after five days to allow pH measurement.

TABLE 2—*Properties of boiling water extracts from polyurethane and phenolic foams* [3].

Properties	Phenolic Foam	Polyurethane Foam
pH free water	2.37	8.4
pH water in foam	2.25	4.3
Water pickup, g	154	60

TABLE 3—*Physical properties of aqueous extracts from various thermal insulations* [3].

	Temperature, $\mu\Omega$[a]			Halogens, ppm	
Insulations	RT	120°F	210°F	Cl^-	Br^-
Polyurethane (FP)	30	45	100	20	390
Polyurethane	40	50	400	...	...
Calcium silicate	200	350	450	...	...
Mineral wool	75	700	...	...	...
Cellular glass	40	60	300	...	...
Fiberglass	220	850	1200	...	...
Ceramic blanket	25	40	100	...	...

[a] $t°C = (t°F - 32)/1.8$. RT is room temperature.

It was pointed out in Paper 1 that excessive concern about the chloride content of the lagging material per se in relation to the risk of stress corrosion cracking of austenitic stainless steels avoids to some extent the real problem, which is the total chloride available to the lagged surface during its lifetime. Chloride from all sources must be considered, including that accumulated during storage and installation, and that available from the atmosphere, rainwater or wash water, and other liquid contaminants in service. Some interesting data from the nuclear industry were quoted and are presented in Table 4. There was little recognition of this problem in lagging system specifications, which tended to concentrate on the lagging material chloride content. Paper 1 reported the following references to chloride content in a sample of 20 specifications from the oil, petrochemical, marine, and power generation industries in the United Kingdom:

1 specified less than 6 ppm
2 specified less than 10 ppm
1 specified less than 20 ppm
1 specified less than 50 ppm
2 specified "low chloride content"
13 gave no specific limit

The physical properties of the various lagging systems in relation to fluid transport were also discussed. There was a consensus that calcium silicate has unfavorable "wicking" properties, and that closed cell foam glass is relatively impermeable. There were differing views on the degree of permeability of organic foams. Regardless of the intrinsic permeability of the lagging material, the role of joints in relation to the permeability of the lagging system was emphasized.

Corrosion Prevention: Inhibited Laggings

The basic technology supporting the use of inhibited lagging to prevent stress corrosion cracking of stainless steels was covered in Papers 1 and 7. Figure 2 is taken from Paper 7 and shows the well known Karnes [*2*] graph, together with some superimposed constraints applied by three American specifiers. Proponents of this approach to corrosion prevention argued the benefits of a relatively high absolute sodium plus silicate level in preventing cracking. Such materials are more tolerant to additional chloride ingress from extraneous sources than those with lower absolute sodium plus silicate levels, which although initially well within the acceptable Karnes' limits, are readily affected adversely by small additional quantities of chloride.

Some contributors expressed concern about exclusive reliance on water soluble inhibitors to prevent cracking. Over the lifetime of a lagging system, progressive water ingress can obviously deplete inhibitor levels below those necessary for cracking prevention. It was pointed out in Paper 1 that stress

TABLE 4—*Measured/available surface contamination levels of chloride on stainless steel foil* [4].

Condition	Chloride Level, $g \cdot m^{-2}$
As supplied	0.0002 (measured)
6 months exposure in workshops	0.0003 (measured)
Finger marked during handling	0.0011 (measured)
Covered with finger prints	0.0054 (measured
Bead of perspiration containing 1000-ppm Cl^-	approximately 1.0 (available)
100-mm insulation, density 100 $kg \cdot m^{-3}$, containing 100 ppm Cl^-	approximately 1.0 (available)

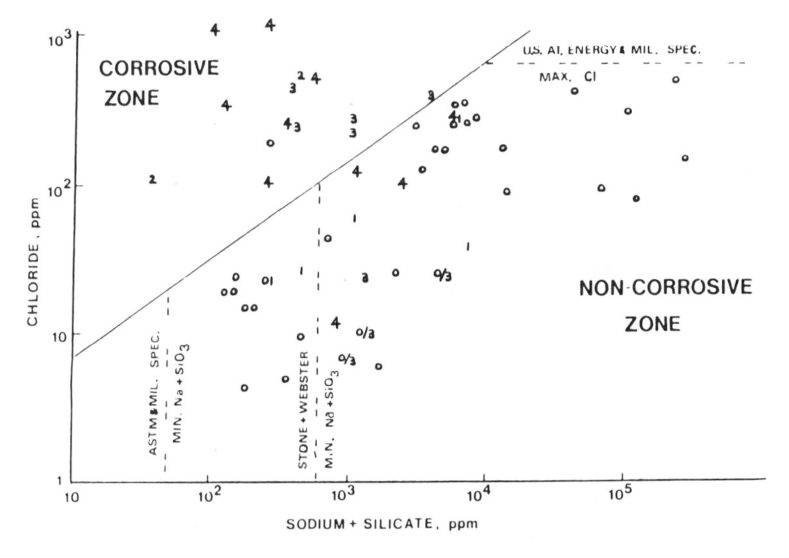

FIG. 2—*Karnes' data* [2] *relating incidence of cracking to silicate/chloride ratio of lagging material. Numbers denote samples cracked out of four, except where stated. Three typical specification constraints are indicated. Presented originally in Paper 7.*

corrosion cracking has been experienced on stainless steel equipment insulated with amosite asbestos bonded with as much as 20% sodium silicate, that is, well above the sodium plus silicate to chloride ratio considered safe in terms of the Karnes' criteria.

No experience was reported on the direct application of sodium silicate to stainless steel surfaces before lagging application, a practice which is apparently favored in the United States.

It was emphasized that inhibited laggings had been developed with the specific objective of combatting stress corrosion cracking of stainless steels. No claims were made or any experience reported of their efficiency in relation to general corrosion of carbon steel under laggings.

Corrosion Prevention: Organic Coatings

The scope for organic coatings in preventing corrosion was covered specifically in Papers 1, 4, 5, and 8.

Regarding carbon steel corrosion, there was a consensus that it is beneficial to apply coatings beneath fireproofing, and beneath lagging systems on steel surfaces operating within the susceptible temperature range cited above. Generally practice appeared to favor the use of relatively simple finishing systems applied to wire brushed surfaces, and among the coatings cited were red lead, zinc chromate, and epoxy or phenolic priming systems, dependent on upper operational temperature. However, in some cases, more sophisticated systems had obviously been felt appropriate, and in the specific case of steel in concrete where a bond is required, the use of epoxy based systems was favored. Paper 8 also outlined the benefits of coating the concrete itself to minimize water diffusion, and successful experiences with water-based vinyl copolymers and acrylic emulsions were cited.

A number of operators cited their use of coatings to control external stress corrosion cracking of austenitic stainless steels. Favored coatings were those with relatively good high temperature properties, and included silicone-alkyds and aluminum filled silicones. Some interesting data were presented in Paper 4 relating to the protection efficiency of a number of commonly specified coating systems, and Table 5 is taken from the paper. The data confirm that relatively cheap systems using single coats of paint on degreased surfaces markedly reduced the incidence of cracking relative to bare surfaces, but are only as efficient as they are free of holidays and impervious. In additional tests, the silicon-alkyd paint proved no more efficient when cured before immersion than when uncured. Evidently no galvanic benefit derives from the use of aluminum-filled silicone paint. One case was reported concerning the

TABLE 5—*Incidence of stress corrosion cracking on coiled 304 spring specimens in boiling saturated sodium chloride solution at 108° C.*[a]

Protection System	Corrosion Potential, mV/SCE[b]	Total Number of Cracks 4 Specimens	Protection Efficiency, %
None (control)	−380	75	...
Silicone-alkyd paint, uncured	−140	8	89
Aluminum-rich silicone paint	−390	8	89
Zinc-rich epoxy paint	−720	2	97
Aluminum foil	−910	0	100

[a] Presented originally in Paper 4, see Appendix.

[b] Potentials recorded at the test temperature of 108°C. SCE is saturated colomel electrode.

successful use of an epoxy paint system to control stress corrosion cracking experienced during coastal site storage in the Middle East.

Stress Corrosion Cracking Prevention: Metallic Foils/Paints

The use of foils, typically 46 standard wire gage (swg), aluminum foil beneath lagging systems to prevent stress corrosion cracking of stainless steels was covered in Paper 4. The foil apparently acts as a physical barrier to the migration of small quantities of aggressive fluid towards stainless steel surfaces, and provides cathodic protection in "flooded" lagging systems, preventing pitting/cracking initiation. Table 5 from the paper summarizes some laboratory data that confirm the galvanic protection afforded by the foil, and that it is more efficient than single-coat paint systems in reducing the risk of stress corrosion crack initiation. The foil can be used on surfaces operating at temperatures up to 500°C and is applied by simply wrapping around pipes or vessels with overlays arranged to shed water.

The use of stainless steel foil as an alternative to aluminum foil was also reported. It has the advantage of being usable at temperatures >500°C, but acts strictly as a physical barrier, and can provide no galvanic protection. One operator reported using soft iron foil at temperatures >500°C, which can provide some galvanic protection in the event of flooding of the lagging system at lower temperatures. Both stainless steel and soft iron are more difficult to apply than aluminum foil.

Paper 1 presented some data concerning the preferences of individual operators for foil or paint coatings to prevent stress corrosion cracking. The 20 specifications referred to above yielded the following:

7 specified paint coatings
5 specified foil
3 specified either paint or foil
5 made no reference

The same supplier also observed that in their experience, approximately 90% of stainless steel surfaces are protected with foil, and approximately 10% with paint coatings.

Two major concerns regarding the use of aluminum foil were voiced in discussion. The first concerned the corrosion resistances of the foil, which is <5 mil thick. Both Papers 4 and 7 reported maintained protection efficiencies despite a considerable degree of perforation of the foil, although it was recognized that prolonged flooding could result in virtual removal of the foil. The second concern related to the risk of liquid metal embrittlement in the event of fire. This was also a major concern in relation to the use of zinc-rich coatings, which are known to be efficient at preventing stress corrosion cracking caused by the galvanic protection imparted to the substrate, as indicated in Table 5.

The general problem of liquid metal embrittlement in relation to the use of zinc and aluminum was addressed in Paper 6. The cracking susceptibilities of a range of ferritic, austenitic-ferritic, and fully austenitic materials had been determined under tensile loading when coated with zinc or aluminum at temperatures up to approximately 850°C, or when welded. Some of the data from the paper are presented in Tables 6 and 7. All of the test materials proved susceptible to embrittlement by zinc, in particular the austenitic materials where a nickel leaching mechanism operates. However, none of the test materials cracked in the presence of aluminum, although in some cases, there was evidence that alloying had occurred. Discussion revealed a consensus that it is inadvisable to coat stainless steels with zinc-rich paints where toxic or flammable materials are being processed. However, embrittlement or cracking risks are significantly lower in the case of aluminum, albeit some alloying might occur, and require detection, in the event of a fire.

Stress Corrosion Cracking Prevention: Materials Fabrication/Selection

A number of cases of stress corrosion cracking reported at the meeting had been caused by the ingress of fluids into laggings, not from the exterior, but from the vessels or piping systems themselves. Typical fabrication defects that had led to flooding of lagging systems included lack of fusion, porosity, and piping, particularly in tack and stitch welds used in the assembly of tanks.

TABLE 6—*Incidence of liquid metal embrittlement of stainless steels and nickel alloys by zinc and aluminum for burner experiments.*[a]

	Fracture[b]	
Material	Zinc	Aluminum
ASTM A 285C	+	
SAE 4140	+	
ASTM A-200T4	+	
5 Cr 0.5 Mo	+	
7 Cr 0.5 Mo	+	
9 Ni	+	
AISI 405	+	−
18 Cr 2 Mo	+	−
26 Cr 1 Mo	+	−
AF 22	+	−
3 RE 60	+	−
AISI 304		−
AISI 316		−
Incoloy 800	+	
Hastelloy C	+	

[a]Presented originally in Paper 6.
[b]Blank means not tested.

TABLE 7—*Incidence of liquid metal embrittlement of stainless steel and nickel alloys by zinc and aluminum for welding experiments.*[a]

Material	Cracks	
	Zinc	Aluminum[b]
AISI 304	+	−
AISI 310	+	
AISI 316	+	−
AISI 317	+	
AISI 321	+	
AISI 347	+	
Incoloy 800	−	
Incoloy 825	+	
Hastelloy B	+	
Hastelloy C	−	
AF 22	−	−
18 Cr 2 Mo	−	
26 Cr 1 Mo	−	−
AISI 405	−	−

[a] Presented originally in Paper 6.
[b] Blank means not tested.

The lessons in relation to appropriate construction supervision and inspection are obvious.

In a number of instances, experience of expensive stress corrosion cracking problems with austenitic stainless steels had resulted in the selection of more resistant alloys for replacement vessels and piping systems. The materials that were actively discussed at the meeting were as follows:

1. *Extra Low Interstitial Ferritic Steels*—The use of 18 Cr 2 Mo grades of ferritic stainless steel for vessels and piping systems in the brewery industry was discussed in Papers 2 and 3. These materials are immune to chloride stress corrosion cracking and can be used at operational temperatures up to approximately 300°C. Concerns were expressed about the use of such materials for welded constructions with wall thicknesses greater than "a few millimetres" because of the problems of achieving adequate heat-affected zone (HAZ) toughnesses. However, Paper 3 reported on the satisfactory construction of some sizeable vessels with wall thicknesses up to 6 mm.
2. *Duplex Stainless Steels*—In a number of cases, duplex 18 Cr 5 Ni grades of stainless steel had been used to construct sizeable vessels. These materials are not immune to stress corrosion cracking but are significantly more resistant in terms of tolerable temperatures and chloride levels than the conventional 18 Cr 8 Ni austenitic grades of stainless steel.
3. *"Super" Austenitic Stainless Steels*—There was some discussion of the scope for 20 Cr 25 Ni grades of austenitic stainless steel, which although not immune to SCC are significantly more resistant than 18 Cr 8 Ni grades. The

consensus was that, in terms of limiting temperature, they perform similarly to the duplex grades, although there was some divergence as to values of that temperature within the range of approximately 140 to 180°C.

Corrosion Prevention: Design, Specification, Inspection, and Maintenance

A recurring theme throughout the meeting was that there are two requirements for controlling corrosion within lagging and fireproofing systems. On the assumption that water can enter the system, there is a need for some anticorrosion measure(s) to be adopted within the system, be it inhibitors, paints, foils, or whatever. However, there is an overriding need to keep water out of lagging and fireproofing systems at all stages from application to retirement. None of the available anticorrosion measures were designed for, or are able to cope with, prolonged periods of exposure to flooded systems, regardless of the lack of efficient insulation offered by such systems. It follows that designs and specifications, while concerned with specific anticorrosion procedures, must also be preoccupied with the necessity for efficient waterproofing, and that this emphasis must be maintained through application or in-service inspection and maintenance.

There was a consensus on the desirability of consultation between designer, operator, and application contractor at the design stage to produce a suitable specification, which was elaborated in Paper 10. In relation to waterproofing, a number of specific mechanical design issues were raised and are worthy of note:

1. "Top hats" are of considerable value in shedding water away from upper termination joints between fireproofing and steel.
2. Joints in metal foil and cladding should always be arranged to shed water.
3. Drainage points should be arranged at the base of insulation systems on long vertical pipe runs, columns, and so forth to prevent water holdup.
4. Waterproof sealant should be applied around any protrusions from lagging systems, such as hangers, supports, and so forth, and these should be kept to a minimum.
5. Joints in steam tracing pipework should always be outside, preferably beneath, the main lagging system.

While there was general recognition of the need for an appropriate nonpermeable vapor barrier on cold insulation systems, there was some divergence as to the relative merits of metal cladding versus reinforced mastic coatings for weatherproofing hot insulation systems.

The case for using "specialist" insulation application inspection was presented in Paper 9. The key tasks for such inspection were identified as follows:

(1) confirmation that correct specified materials are applied in each area,
(2) approval or control of storage to avoid wetting,
(3) confirmation that insulant remains dry after application until sealing or cladding is completed,
(4) checking correct application of vapor barrier,
(5) checking correct cladding application, including arrangement of joints to shed water, and sealing of gaps, cut-outs, and so forth,
(6) confirming appropriate staggering of joints in multilayer insulation systems,
(7) checking for damage at all stages of application, and
(8) witnessing laboratory tests on in-situ foamed materials.

There was also a consensus that vigilance on many of the latter points needs to be maintained throughout the life of the lagging system. Caulking or jointing materials dry out and lose flexibility, cladding or barrier systems suffer local damage or perforation, joints leak in service and so forth. Accepting that the corrosion control system within the lagging or fireproofing system cannot provide unlimited containment, such problems need to be identified and remedied, or corrosion problems are inevitable.

Finally, the in-service inspection of metal surfaces beneath lagging systems was discussed briefly. The specific use of a magnetoscope for detecting zinc embrittlement of austenitic stainless steel surfaces was discussed in Paper 6. Otherwise, no new initiatives were reported at the meeting, and there appeared little alternative to the costly removal of lagging/fireproofing systems to allow access for the traditional nondestructive testing (NDT) techniques. There was a consensus that this is an unsatisfactory situation, and that there is a need for the development of an appropriate in-situ technique.

APPENDIX

Abstracts of Papers Presented at the Corrosion under Lagging Conference Held in Newcastle-on-Tyne, England, Nov. 1980

Paper 1

"Thermal Insulation. Specification and Materials. Application to Stainless Steel Surfaces," J. D. Nicholson, Darlington Insulation Co Ltd., West Auckland Road, Darlington, Co Durham DL3 OUP, United Kingdom.

The problem of external stress corrosion cracking of certain grades of stainless steel is well known in the process industries. Soluble chlorides and fluorides in contact with austenitic stainless steels under stressed conditions create a potential risk, which can be reduced if the plant designer chooses the optimum grade of steel.

Insulating materials are available which, in association with suitable accessories, can reduce the risk of stress corrosion cracking. Materials with low content of free chloride, the presence of barrier materials, and inhibited insulation can be of value.

Reproducibility of results is difficult under laboratory conditions. Site experience suggests that dual temperature operation may provide greater risk than does continuous high temperature operation.

A review of British Standards and of specifications that are issued by engineering and petrochemical companies shows divergence of opinion about permissible levels of soluble chloride.

The insulation contractor can assist in preventing dangerous conditions by the correct design and application of insulation systems.

Inspection and regular maintenance are essential to preserve the integrity of an insulation system. The risk of stress corrosion attack can be reduced but not necessarily eliminated by the use of materials of low chloride content.

There is only one way to prevent External Stress Corrosion Cracking of stainless steel in contact with insulation: Keep the Insulation Dry at All Times.

This paper was published subsequently in the *Bulletin of the Institution of Corrosion Science and Technology*, Vol. 19, No. 5, Oct. 1981, pp. 2-5.

Paper 2

"The Basics of Corrosion Mechanisms in Lagged Steelwork," RAE Hooper, Stainless Steels Department, BSC Sheffield Laboratories, Swindon House, Moorgate, Rotherham S60 3AR, United Kingdom.

Process plant vessels and pipework are usually lagged to conserve heat, and occasionally cold, and this has been a common practice for a very long time. The sharp rise in fuel prices in the last decade, and the other increasing need to conserve energy, is making the lagging of process equipment more and more necessary. The usual effect of lagging is to raise the temperature of the outer side of the plant wall, which is most commonly made of steel. This should be dry and free from corrosion hazards but, unfortunately, this ideal situation frequently does not occur as moisture enters the lagging. On some occasions this arises from undetected leaks in the process equipment but more often the moisture comes in through inadequate weather protection of the lagging. Lagging should be covered with a waterproof coating to provide protection from the rain, sea spray, or spillage or leakage from other parts of the plant. When moisture does get into the lagging the high temperature causes evaporation and concentration of any dissolved solids, either in the incoming water or leached from the lagging itself. The point of maximum concentration depends upon the rate of water ingress, the plant temperature, and the extent of temperature transients, but it is inevitable that the hot plant wall will see an environment concentrated in the dissolved solids. As a result, corrosion can occur.

Carbon and low alloy steels tend to suffer general attack in hot, moist environments but if the liquids trapped under lagging contain substances such as nitrates or caustic, then stress corrosion cracking can also occur. This type of corrosion is typified by intergranular cracks and can be rapid under some circumstances. Stainless steels generally suffer from pitting or crevice corrosion rather than general corrosion, and the most harmful dissolved species are the halogens, of which chlorides are by far the most common and most aggressive. If the stainless steel has an austenitic structure, for example, Types 304 and 316 (Unified Numbering System [UNS]), which contain approximately 10% nickel, then stress corrosion cracking can also occur in the presence of chlorides. This type of cracking is characteristically transgranular and highly branched, although in some instances some intergranular cracking can also occur.

Corrosion under lagging can be prevented by a number of methods, which include the prevention of ingress of moisture, the use of laggings which do not contain harmful, leachable chemical species, the surface coating of pipes and vessels and last, but by no means least, the correct choice of material.

Paper 3

"External Use of Austenitic Stainless Steel Vessels and Pipework—A Case Study," D. Geary and G. Bailey, CAPCIS, Corrosion and Protection Centre, UMIST, P.O. Box 88, Manchester M60 IQD, United Kingdom.

No abstract supplied.

Paper 4

"ICI Practice for Preventing SCC Under Lagging," M. E. D. Turner, Consultant, 1 Norfolk Crescent, Ormesby, Cleveland TS3 OLY, United Kingdom.

Experience has shown that a significant reduction in the risk of stress corrosion of stainless steel under lagging can be effected by the simple expedient of interposing a layer of aluminum foil between the equipment and the lagging.

The underlying theory is explained and laboratory test and practical experience described.

Some alternatives are discussed and the reasons for their rejection given.

Paper 5

"Use of Paint Coatings for Under-Lagging Corrosion Prevention. A User View," F. H. Palmer, Engineering Department, BP Trading Ltd., Britannic House, Moorgate, London EC3, United Kingdom.

No abstract supplied.

Paper 6

"Stress Induced Cracking of Steels by Molten Zinc and Aluminum," P. Geenen, Koninklijke/Shell-Laboratorium, Shell Research B.V., Postbus 3003, 1003AA, Amsterdam, Netherlands.

After the Flixborough disaster had highlighted the risk of contact between molten zinc and austenitic stainless steels, failures caused by such contact have been reported regularly. Disagreement in the literature about the prevailing mechanism and ignorance of what materials are susceptible have stimulated our interest in this matter. Particularly the need for a nondestructive detection technique called for an experimental program.

The work was done in cooperation with the laboratory of Inorganic Chemistry and Materials Science, Department of Chemical Engineering, Twente University of Technology, Enschede, Holland.

Specimens of the relevant materials were coated with a zinc or aluminum compound, subsequently TIG welded without filler material and examined microscopically. SEM and X-ray distribution images were made. The susceptibility of a range of alloys has been determined quantitatively by comparing the time to fracture and the maximum loads applied during slow straining of specimens, with and without zinc coating or aluminum metal layer.

A basis for nondestructive testing was laid by comparing observed metal attack under zero strain and measured change in relative magnetic permeability by means of a Forster Magnetoscope.

Paper 7

"An Experimental Investigation with Stress Corrosion Cracking of Austenitic Stainless Steel under Insulation," J. Gillett and K. A. Johnson, Fibreglass Ltd., Insulation Division, St. Helens, Merseyside, WA10 3TR, United Kingdom.

The paper discusses the definition of insulation, its usage and the conditions that may be encountered during application and afterwards in service.

The discussion includes the definition of when corrosion can be a problem and having apparently restricted it to a narrow range of conditions demonstrates that, in fact, these conditions are likely to be accounted transiently even if only infrequently and so may be more common than is generally supposed. The possible origins of chloride ions are listed and discussed.

The paper highlights the current approach of specifiers with the general requirement of low Cl^- being very common. However many international authorities do not follow the U.K. approach, and require high sodium and silicate contents for inhibition purposes.

Investigation of fibreglass materials was followed by an extensive test program to establish whether or not fibreglass materials could cause corrosion. These tests demonstrate that in fact fibreglass materials could cause corrosion, in fact that fibreglass does not cause corrosion of stainless steel. Further, the program investigates protective measures that could be taken against corrosion caused by contamination from external sources of Cl^- ions.

The paper concludes with discussion of the results and recommendations for future specifications.

Printed copies of this paper are available from Fibreglass Ltd.

Paper 8

"Corrosion Control Under Thermal Insulation and Fireproofing," J. F. Delahunt, Exxon Research and Engineering Co., P. O. Box 101, Florham Park, NJ 07932.

In recent years there have been an increasing number of reports concerning corrosion occurring on carbon steel structures and equipment that are either thermally insulated for energy conservation or coated with concrete for protection from fire. In view of this, Exxon Engineering has been involved in a number of field investigations as well as laboratory investigative programs to evaluate the cause of corrosion and to determine appropriate means to mitigate it. The cumulative result of these various programs is presented within this paper, and it includes discussions concerning:

- Examples of corrosion in refineries, petrochemical plants, and pipelines occurring on insulated or fireproofed structures and equipment.
- Descriptions of potential corrosion mechanisms.
- Corrosion mitigation systems used to prevent attack of such equipment.

Paper 9

"Application Inspection of Insulating Materials," P. G. Blackburn and R. G. Roberts, ITI Anti-Corrosion Ltd, 177 Hagden Lane, Watford, Hertfordshire WD1 8LW, United Kingdom.

In order to minimize heat losses it is important that thermal insulation is correctly applied. Insulation inspectors, operating independently of the contractors and on be-

half of the client, can play an important part in the operation. Particularly in highlighting defects that would subsequently, and very expensively, waste heat or promote corrosion.

Insulation materials should be inspected before installation for any damaged areas. Joints should be offset, slabs or sections correctly butted, and in the case of two layer application the inspector must ensure that horizontal and vertical joints are correctly positioned.

Insulation on bends is a potential weakness because careless cutting on site can leave gaps. These gaps are sometimes masked by the bands used to retain the insulation in position. The bands themselves can also cause damage to fragile insulating materials.

Metal cladding joints have to be correctly overlapped and wrongly positioned joints can encourage water ingress. it is also important to see that joint sealants are correctly applied or that fixing methods, such as pop rivets, do not penetrate the insulating layer.

In general the most competent insulation inspectors are recruited from the thermal insulating industry. Preferably they would have had at least ten years practical experience so that they are well aware of the short cuts, mistakes, and malpractices of operatives.

Paper 10

"How to Write a Specification," I. G. Huggett, Consultant, Orchard House, Low Worsall, Yarm, Cleveland, United Kingdom.

The specification has an important role in the lagging of process plant and is both the interface and the means of communication between the lagging contractor and the client. Thus, for example, whatever decisions are reached about the precautions necessary to minimize corrosion, they have to be defined and put into effect by means of the specification.

A good specification for lagging, as for other relatively complex construction activities, is hard to write and even the best cannot incorporate all that is necessary to achieve good lagging; thus it can never be completely self-contained. Some things have to be planned in advance by the client or others; additional communication between client and contractor is almost always worthwhile and rarely unnecessary.

The way in which the specification is drafted, the phraseology, and the way the ideas are expressed are always important. Frequently there is also real technical difficulty in coming to decisions, and great care is necessary when trying to formulate written rules: the provision for minimizing corrosion often falls into this category.

References

[1] Page, G. G., *New Zealand Journal of Dairy Science and Technology*, Vol. 15, 1980, pp. 143-157.

[2] Karnes, H. F., American Institute of Chemical Engineers 57th National Meeting, Sept. 1965.

[3] Delahunt, J. F. *Insulation Journal*, Vol. 26, No. 2, Feb. 1982, p. 10.

[4] Nicholson, J. D., *Bulletin of the Institution of Corrosion Science and Technology*, Vol. 19, No. 5, Oct. 1981, p. 2.

Thermal Insulation Materials

George E. Lang[1]

Thermal Insulation Materials: Generic Types and Their Properties

---

REFERENCE: Lang, G. E., **"Thermal Insulation Materials: Generic Types and Their Properties,"** *Corrosion of Metals Under Thermal Insulation,* "*ASTM STP 880*, W. I. Pollock and J. M. Barnhart, Eds., American Society for Testing and Materials, Philadelphia, 1985, pp. 63–68.

ABSTRACT: Thermal insulation plays a very important function in all of our lives. Without it the cost of energy used in our homes would skyrocket, perishable foods would be available only within a short distance of the source, plastic items and many petroleum products would not exist, and we would have to get used to warm beer and soupy ice cream.

Now to the generic and functional differences of the various types of thermal insulation. One of the most defining properties of a thermal insulation is its service temperature range. Since we are discussing generic types of insulation, and some manufacturers' recommended service temperatures vary slightly, the du Pont Company's established recommended service temperatures are listed in the paper. The properties, good and bad, of generic thermal insulation materials that have a bearing on their individual influence on corrosion of the metals they cover will be discussed.

KEY WORDS: thermal insulation, polystyrene, plastics, corrosion

Thermal insulation plays a very important function in all of our lives. Without it the cost of energy used in our homes would skyrocket, perishable foods would be available only within a short distance of the source, plastic items and many petroleum products would not exist, and we would have to get used to warm beer and soupy ice cream.

Except for our concerns about home heating costs, thermal insulation is not appreciated in the remote circumstances where it influences our lives, and in very many of these cases it is neglected.

Thermal insulation, in the form of pipe covering, block, batt, and blanket, is many small, to microscopically small, gas-filled cells of some form of solid material. The material may be foamed plastic or glass, it may be a block or blanket of glass or mineral fibers, or it may be a cementitious mixture known as calcium silicate or perlite silicate.

[1]Consultant on thermal insulation, E. I. du Pont de Nemours and Co., Inc., Engineering Service Division, Louviers Building, Wilmington, DE 19898.

One of the most defining properties of a thermal insulation is its service temperature range. The established service temperatures recommended by the du Pont Company's engineering department for the thermal insulation materials it normally uses, as generic types are shown in Table 1.

The properties and characteristics of these generic thermal insulation materials, which have a bearing on their individual influence on corrosion of the metals they cover, are discussed in the following sections. The information is based on our many years of experience.

I have not included comments on thermal efficiency. The reason is that manufacturers' literature values are for thermal conductivities determined in laboratories for "dry" samples. But installed thermal insulation frequently contains more moisture than these samples. This increased moisture may have come from shipping, storage, installation, or service. The good low K values of thermal insulations are quickly lost when the insulations become wet. Therefore, in the field we frequently have poorer thermal efficiency of the installed insulation than the K values used in the initial design calculations. In the practical sense, comparing thermal insulations solely on the K values from "dry" samples can be misleading. What their values are in the field when they have more water in them depends, of course, on the amount of water. When thoroughly wet, many have similar high K values (Table 2).

Polystyrene Foam

• This insulation is available in the form of polystyrene foam billets or polystyrene beads expanded in a mold.

• To be functional as a process thermal insulation, the density cannot be less than 24 kg/m^3 (1.5 lb/ft^3).

• In the prescribed density, it does not absorb or wick water as long as the cell structure remains intact.

TABLE 1—*Established thermal temperatures recommended by the du Pont Company's engineering department.*

Generic Thermal Insulation Materials	Recommended Service Temperature, °C (°F)	
Polystyrene foam	−73 to 60	(−100 to 140)
Polyurethane foam—rigid	−73 to 82	(−100 to 180)
Polyisocyanurate—rigid	−73 to 149	(−100 to 300)
Flexible foamed elastomer	+2 to 82	(+35 to 180)
Cellular glass	−129 to 149	(−200 to 300)
Glass fiber	4 to 190 or 454 depending on type	(40 to 375 or 850)
Mineral wool	60 to 649 or 982 depending on type	(140 to 1200 or 1800)
Calcium silicate	60 to 649	(140 to 1200)
Perlite silicate	60 to 593	(140 to 1100)

TABLE 2—*Comparison of* K *values from samples in the field.*

Samples	K BTU, in./h, °F, ft^2
Dry calcium silicate	about 0.42
Dry urethane	" 0.20
Dry fiberglass	" 0.25
Thoroughly wet calcium silicate	" 4.0
Thoroughly wet urethane	" 4.0
Thoroughly wet fiberglass	" 4.0

- It has a maximum service temperature of 60°C (140°F).
- Compressive resistance is 172 kPa (25 psi) average at 5% deformation.
- The insulation is destroyed by solvents other than alcohol and is softened under black or other final coverings that will reach 60°C (140°F) or above.
- We have not used a significant quantity of this type insulation. No plant corrosion problems have been noted.

Polyurethane Foam—Rigid

- This plastic foam is primarily used for cold and anti-sweat service.
- It does not absorb and wick water as long as the cell structure remains intact. It is permeable to water vapor in cold service when required vapor barrier fails. Vapor diffuses through cell walls to the temperature zone where it condenses and further to where it freezes.
- It has a maximum service temperature of 82°C (180°F).
- The typical average bulk density is 32 kg/m^3 (2 lb/ft^3).
- Compressive resistance is 17 kPa (25 psi) average at 5% deformation.
- If in continuously cold service, it does not corrode unprotected metal surfaces. If in intermittent service to its maximum service temperature, it can cause corrosion of unprotected wet metal surfaces from released chlorides in fire retardants and blowing agents. The sun's ultraviolet rays decompose this insulation.

Polyisocyanurate Foam—Rigid

- This insulation is very fire resistant for an organic foam; it is a low flame propagation rate plastic foam of the polyurethane family.
- It does not absorb and wick water as long as the cell structure remains intact. It is permeable to water vapor in cold service when required vapor barriers fail.
- Maximum service temperature is 149°C (300°F).
- Typical average bulk density is 32 kg/m^3 (2 lb/ft^3).
- Compressive resistance is 17 kPa (25 psi) average at 5% deformation.

• When this material is exposed to heat and moisture, the cell structure in the heated zone is damaged. The decomposition products may contain chlorides from the fire retardant and blowing agent and are thus fairly aggressive as corrodants of unprotected wet metal surfaces. The sun's ultraviolet rays decompose this insulation.

Flexible Foamed Elastomer

• This insulation is a black, rubbery type flexible foamed plastic.
• It does not readily absorb or wick water.
• The maximum service temperature is 82°C (180°F).
• Average bulk density is 96 kg/m^3 (6 lb/ft^3).
• It has poor compressive resistance.
• Although not corrosive by itself, it supports corrosion of unprotected metal surfaces when water is present, particularly when the water contains chlorides from an external source.

Cellular Glass

• This insulation is a rigid glass foam whose blowing agent contains hydrogen sulfide and carbon dioxide.
• It does not absorb and wick water as long as the cell structure remains intact.
• The maximum service temperature is 149°C (300°F).
• Average bulk density is 136 kg/m^3 (8.5 lb/ft^3)
• Compressive resistance is 690kPa (100 psi) average.
• When water is present and the cell structure is damaged, release of the foam blowing agent may cause corrosion on unprotected carbon steel surfaces.

Glass Fiber

• This insulation is pure glass fiber containing various types of binders.
• It will absorb and wick water, but is more able to drain excess moisture than other types of insulation.
• The maximum service temperature is 191 to 454°C (375 to 850°F) depending upon type and brand.
• The bulk densities range from 24 to 96 kg/m^3 (1.5 to 6 lb/ft^3) depending upon type and brand.
• It has poor compressive resistance.
• The fact that it will wick water makes it conducive to corrosion on unprotected wet metal surfaces.

Mineral Wool

• This insulation is a mineral or metal slag fiber, basically an impure glass fiber.

- It readily absorbs and wicks water.
- Maximum service temperature is 649 to 982°C (1200 to 1800°F) depending upon brand.
- Bulk densities range from 96 to 304 kg/m^3 (6 to 19 lb/ft^3) depending upon type and brand.
- It has poor compressive resistance; there is some improvement at higher densities.
- The fact that it will wick and hold water makes it conducive to corrosion on unprotected wet metal surfaces.

Calcium Silicate

- This insulation is a cementitious mixture, known in the trade as "white goods."
- It readily absorbs and wicks water. Can hold up to 400% of its own weight of water without dripping.
- Maximum service temperature is 649°C (1200°F).
- Average bulk density is approximately 224 kg/m^3 (14 lb/ft^3).
- Compressive resistance is 621 to 1103 kPa (90 to 160 psi) at 5% deformation, depending upon brand.
- Although its pH is initially high, 10 average, it is fairly aggressive in supporting corrosion on unprotected wet metal surfaces because of its moisture retention, particularly when the moisture contains chlorides from an external source.

Perlite-Silicate

- This insulation consists mostly of expanded perlite with sodium silicate as a binder.
- It may contain an ingredient to resist water absorption up to a certain temperature at which point the moisture resistance burns out. When this happens, it will absorb and wick water.
- The maximum service temperature is 593°C (1100°F), above which shrinkage becomes excessive.
- Average bulk density is approximately 224 kg/m^3 (14 lb/ft^3).
- Compressive resistance is somewhat less than the calcium silicates being in the 483 to 552 kPa (70 to 80 psi) range at 5% deformation.
- The pH is high, being 10 plus, making it less conducive as a corrodant on wet metal surfaces particularly when it contains a water repellant.

Conclusion

Selection of a thermal insulation depends on many factors. We believe an important characteristic to consider is what extent the insulation can contribute to corrosion of the metal underneath if the insulation gets wet. In general, we prefer "sterile" insulations. However, even a totally sterile insulation will accelerate the corrosion process going on under it on an unprotected wet metal surface.

DISCUSSION

W. D. Johnston[1] *(written discussion)*—I wish to comment on Mr. Lang's reference to cellular glass. Mr. Lang indicates that cellular glass provides a source of sulfuric acid that is aggressively corrosive. This is not true. FOAMGLAS® cellular glass insulation is not a source of sulfuric acid. In fact, under aggressive hydrolysis conditions, it yields an alkaline solution of about pH = 9, which is inconsistent with the presence of sulfuric acid.

G. E. Lang (author's closure)—My entire paper dealt primarily with corrosion experiences du Pont has had on its industrial plants. Those experiences over 20 plus years have been accumulated by a number of engineering people in addition to Roy Allen and myself. These plant experiences are confirmed by our laboratory tests where possible.

As we heard during the symposium, severe corrosion and or failures normally take years to occur. For this reason, a leachant prepared in a ½-h test does not go far enough to guide us in solving our "real world" corrosion problems.

From our actual field experience on a number of vessels, I must stand behind my statement that "cellular glass is very aggressive on unprotected carbon steel surfaces" when in the presence of stagnant warm water.

Please note that in my paper, or discussion with you during the question period or our one-on-one discussion after the meeting, I did not state that FOAMGLAS causes chloride stress-corrosion cracking (SCC) of austenitic stainless steel. In fact, in four 28-day tests over the years, we have proved FOAMGLAS does not cause chloride SCC.

In the few field experiences where we had this SCC problem under FOAMGLAS, we proved the chlorides did not come from the FOAMGLAS.

To update our mutual concern about the pH of a FOAMGLAS leachant, I have set up a test to repeat essentially the Lehigh Lab procedure and Roy Allen's test. This test will run for one year with a monthly check of the leachant. We will keep you advised of the results.

[1] Pittsburgh Corning Corp., 800 Presque Isle Drive, Pittsburgh, PA 15239.

Field Experience

Tore Sandberg[1]

Experience with Corrosion Beneath Thermal Insulation in a Petrochemical Plant

REFERENCE: Sandberg, T., "**Experience with Corrosion Beneath Thermal Insulation in a Petrochemical Plant,**" *Corrosion of Metals Under Thermal Insulation, ASTM STP 880*, W. I. Pollock and J. M. Barnhart, Eds., American Society for Testing and Materials, Philadelphia, 1985, pp. 71-85.

ABSTRACT: A program to detect corrosion under insulation has been ongoing at the Stenungsund ethylene plant since 1976. Experience to prioritize inspection has been developed. During the years 1978 to 1980, the number of yearly inspected lines with more than 0.5-mm corrosion decreased from 50 to 15%. A survey of the condition of 530 pressure vessels was carried out in 1980. Corrosion was found under hot insulation on 75% of not painted vessels and on 20% of painted and insulated vessels. No corrosion was found on equipment in continuous cold service.

KEY WORDS: corrosion, atmospheric corrosion, insulation, experience

The Esso Chemical AB ethylene plant is located in Stenungsund on the west coast of Sweden. The plant was built in 1962 and 1963 with a major expansion in 1969. The plant capacity is 350.000 tons ethylene/year.

In 1976 severe corrosion under insulation (CUI) was found on piping, and a couple of years later major corrosion was also found on other equipment. At that time a program was started to detect corrosion and make required repairs.

In 1980 severe corrosion on two towers caused a plant shutdown. Extensive stripping of equipment was carried out to complete the corrosion picture.

Several actions have been taken to prevent future corrosion, such as

(1) increased resources for inspection and maintenance,
(2) new specifications for equipment details, painting, and insulation, and
(3) new demisters were specified for cooling tower installation.

[1]Engineering associate, Esso Chemical AB, Box 852, S-444 01 Stenungsund, Sweden.

Thermal Insulation Systems Used

Insulation for Hot Service

In part of the plant built in 1962 and 1963 calcium silicate insulation was used throughout with jacketing of galvanized and painted steel plate. Process equipment was painted with red lead before insulation.

In part of the plant built after 1963 and in a particular expansion project 1969 mineral wool insulation was used extensively with aluminum jacketing. Most equipment was not painted before insulation.

Insulation for Cold Service

Block polyurethane insulation and foamed in-place polyurethane was used. Jacketing was painted galvanized steel or aluminum.

As a result of corrosion findings the specification has been revised and painting is now performed before insulating. The latest revision of the insulation specification prescribes carbon steel piping and equipment with operating temperatures (continuous and intermittent services) up to 120°C-continuous, and 200°C-intermittent shall be painted and insulated as follows:

1. Surface preparation—Very thorough blast cleaning in accordance with Sa 2½, Swedish Standard (SIS) 055900-1967 (Pictorial surface preparation standards for painting steel surfaces).
2. Undercoat using inorganic zinc silicate to a thickness of 75 μm.
3. Finish coat—Two-pack, high build, polyamide-cured epoxy, 50 μm total thickness.
4. Nominal dry film thickness for total system, 125 μm.
5. Use mineral wool as insulation material. Jacketing is to be carbon steel with an overlay consisting of 55% aluminum, 43.4% zinc, and 1.6% silicon. Aluminum is used where mechanical damage is less likely.

Environmental and Site Conditions

Stenungsund is located close to the sea. In addition, there is a strong effect from the salt water cooling tower. Plant atmospheric corrosion is affected by the cooling tower mainly because of fogging and salt water mist. The corrosion rate is estimated to increase by a factor of 2 to 5 with increasing proximity to the cooling tower.

The local environmental conditions are as follows:

Air temperature, year average, +7°C
  July, +17°C
  January, −1°C
Relative humidity, year average, 78%

Annual precipitation, 700 mm
Sulfur dioxide in air, 0.004 ppm
Salt in air, outside cooling tower area, 15 g/m^2, year
cooling tower area (80 000 m^2, 260 g/m^2, year
close to cooling tower (maximum), 3500 g/m^2, year

Since 1982, after installation of new demisters in the cooling tower, the amount of salt in the air has been less than 30 g/m^2 for the cooling tower area and less than 175 g/m^2 for regions close to the cooling tower.

Corrosion Encountered

During the years 1963 through 1969 corrosion maintenance and corrosion inspection programs were developed. Spot examinations of equipment showed no major concerns regarding CUI. Our inspection program was considered to be built on several years experience and was kept unchanged after the large plant expansion in 1969.

Piping

In February 1976 the first major problems appeared. In an area, close to the cooling tower, severe corrosion was found under insulation. Lines had to be taken out of operation to be repaired.

A number of questions had to be answered:

- Where to look (area, service conditions, type of insulation, and so forth)?
- How to look (technique, organization)?
- How serious were the problems?
- What were the reasons?
- How to prevent corrosion in the future?
- How to prioritize the inspection?
- Costs involved?
- And many more questions.

During March through May 1976, a comprehensive corrosion detection program was carried out on hydrocarbon lines, hot service insulated and with an operating temperature less than 150°C. During the three month period, 317 lines were inspected by removing sections of insulation.

The following are the results from the 1976 spring investigation: About 18% of the lines had areas with more than 1-mm corrosion. For more details see Table 1. It was also noticed that steam-traced lines were corroded to about the same extent as nonsteam-traced lines. Examples of typical CUI are shown on Figs. 1 through 4.

We concluded from these results that an intensive inspection program must be organized. A prioritization of offsite lines and facilities in the 1969 plant

TABLE 1—*Inspection of 317 lines, March through May 1976. All lines hot service insulated. Operating temperature less than 150°C.*

		Corrosion per 100 Lines				
Area	Number of Inspected Lines	Less Than 0.1 mm	0.1 to 1 mm	1 to 2 mm	2 to 3 mm	More Than 3 mm
1963 plant	91	56[a]	41	2	0	1
1969 plant	155	30	51	10	5	4
Offsite	71	20	45	17	13	5
All areas	317	35	47	9	5	4

[a] Example: 56% of 91 lines from 1963 had less than 0.1-mm corrosion.

was made. All high risk lines were to be inspected before the end of 1976. From 1976 to 1981, 1430 lines were inspected. Of these lines repairs were made as listed in Table 2.

We increased our ability to predict corrosion from 1976 to 1978 as a result of the knowledge built up in 1976. After 1978 there was a dramatic decrease in the number of corrosion findings. This is viewed as a result of the experience in how to prioritize an inspection program for corrosion detection. This experience can be summarized with the following list of criteria for selection of lines to be inspected.

Services Most Likely to Have Corrosion Present

- Unpainted carbon steel with insulation
- Operating temperatures from −10 to +150°C
- Cyclic service
- Poor jacketing
- Exposed to moisture (weather or cooling tower)
- Piping design-configuration and details

Corrosion of uninsulated lines was found to be about the same in each year and much less severe when compared with insulated lines. During 1978 through 79, 2.5% of uninsulated lines had at least one spot corroded to half wall thickness. The corresponding figure for insulated lines was 27%. To illustrate the results of the prioritized inspection, two periods are compared in Table 3.

Stress corrosion in American Iron and Steel Institute (AISI) Type 304 and 316 (Unified Numbering System [UNS] S30400 and S31600) stainless steel has been experienced mainly on small diameter instrument piping. These lines are replaced using material that is better resistant to stress corrosion.

FIG. 1—*Examples of typical CUI—trapped water:* (a) *bottom of bend,* (b) *and* (c) *water entrance at vend may cause corrosion at drain. Example of trapped water is also shown on Fig. 8a.*

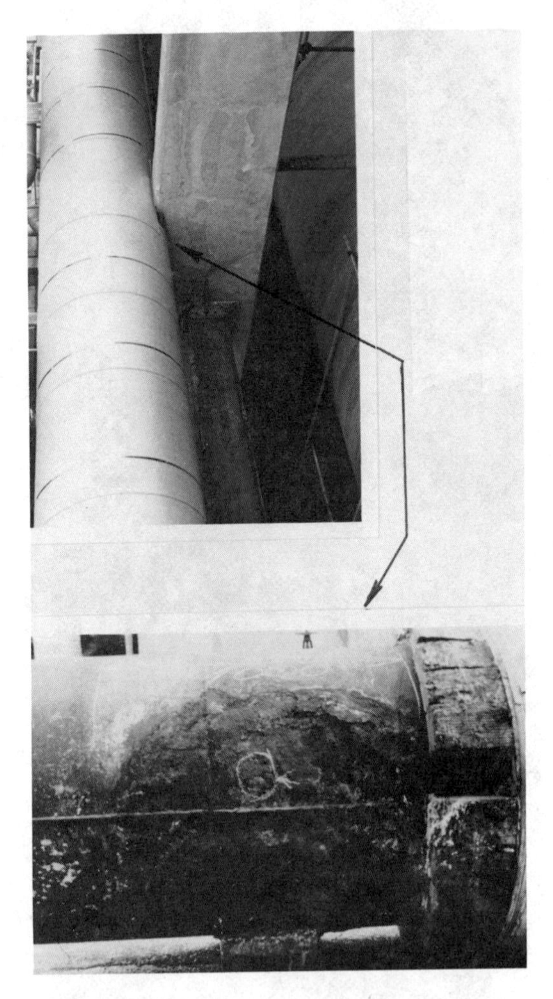

FIG. 2—*Example of CUI—jacketing open at support. Insulation jacket open at vertical beam caused severe local corrosion. The line diameter is 900 mm.*

FIG. 3—*Examples of typical CUI—effect of painting:* (a) *band of shop painting locally prevented CUI and* (b) *inorganic zinc silicate without top coat does not protect for CUI. Insulation removed after four years.*

FIG. 4—*Support of insulated line without pipe shoe.*

TABLE 2—*Inspection of 1.430 lines during the years 1976 through 1981. 70% of the lines are hot service insulated.*

Year	Number of Inspected Lines	Repaired Lines per 100 of Inspected Lines[a]
1976	350	4.3
1977	...	...
1978	240	20.8
1979	250	4.0
1980	310	2.3
1981	280	1.8

[a] Material is carbon steel in all repaired lines.

TABLE 3—*Illustration of the result of prioritized inspection. Hot service insulated lines (Total number 478).*

	Corrosion per 100 Lines	
Period of Inspection	More than 1/2 Wall Thickness	More than 0.5 mm
Jan. 1978 through Sept. 1979	27	50
Oct. 1979 through Dec. 1980	6	15

Pressure Vessels

Routine inspection of pressure vessels focused on internal inspection. Inspection under insulation was carried out on a spot check basis. Significant outside corrosion was first noted in 1978.

In 1980 major corrosion problems were identified. Severe corrosion on one column required the plant to shutdown. An extensive survey of the condition of insulated equipment was carried out.

In the plant there are 530 pressure vessels of which 190 are insulated. Results of the survey are summarized in Table 4.

Important equipment factors are as follows:

- Operating temperature
- Corrosion protection (painting)
- Mechanical design details
- Standard of insulation
- Standard of jacketing

The environmental conditions and the equipment condition explain the results listed in Table 4. What can not be seen in the table is that cyclic temperature service increases the risk of corrosion for hot and cold insulated equipment.

The high extent of corrosion on hot insulated, unpainted equipment is a result of the above factors. The reasons identified for corrosion on painted equipment were ascribed mainly to poor detailed mechanical design.

Inspection Technique Developed and Employed

In a plant, with a large amount of equipment, organization of inspection is as important as the inspection technique. For piping systems it is necessary to have a good line register. The register should include data, such as operating conditions, type of insulation, painting, time for inspection, inspection inter-

TABLE 4—*Corrosion on pressure vessels, result of survey in 1980.*[a]

Category of Vessel	Corrosion per 100 Vessels		
	None	Moderate	Severe
Uninsulated, painted	88	12	0
Cold insulated (polyurethane) continuous service less than minus 10°C	100	...	...
Hot insulated, painted	80	20	0
Hot insulated, not painted	25	63	12

[a] Total number of pressure vessels: 530, hot service insulated: 90, and cold service insulated: 100.

val, findings, and so forth. It is quickly evident that good updated isometric drawings and flow diagrams greatly simplify inspection, prework, and recording.

Using the data in the line register a judgment can be made about the risk of corrosion under insulation. If there is a risk, the inspector can look for the following details:

- Attachments to piping and process equipment
- Connections, especially vertical segments
- Unsealed or damaged jacketing
- Signs of rust
- Low points especially if jacketing is damaged above the low point
- Bottom of absorbent insulation
- Dead legs

Insulation is removed for spot checking. Complete removal is only required for exceptional cases. Inspection areas or openings in the insulation are not recommended. Inspection openings are more likely to cause corrosion than prevent it. Nondestructive testing methods have so far proven to be unreliable. Figure 5 gives an example where critical corrosion areas are likely to be found.

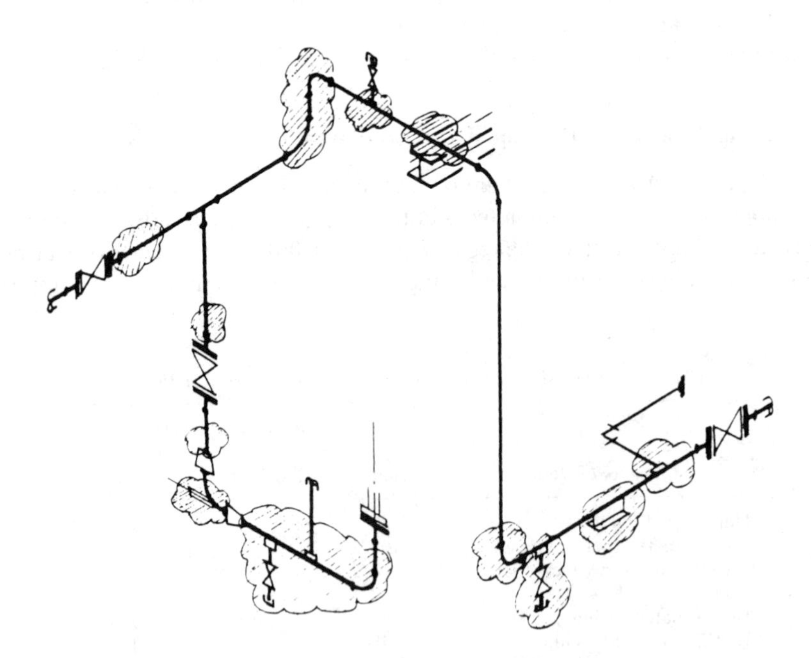

FIG. 5—*Areas critical for CUI.*

In a new plant, inspection of piping should be organized in such a way that all lines in a pipe rack are inspected on the same occasion. If severe CUI is discovered it may be more effective to select the most critical lines and inspect line by line until the situation is under control. When a continuous inspection level is achieved, inspection pipe rack by pipe rack is again preferable.

Our interval for external inspection is three to six years for drums, exchangers, and so forth. For piping our recommendation is six to twelve years. However a first check on piping should be made three to six years from start-up in order to obtain a general impression on piping condition and to set future inspection timing.

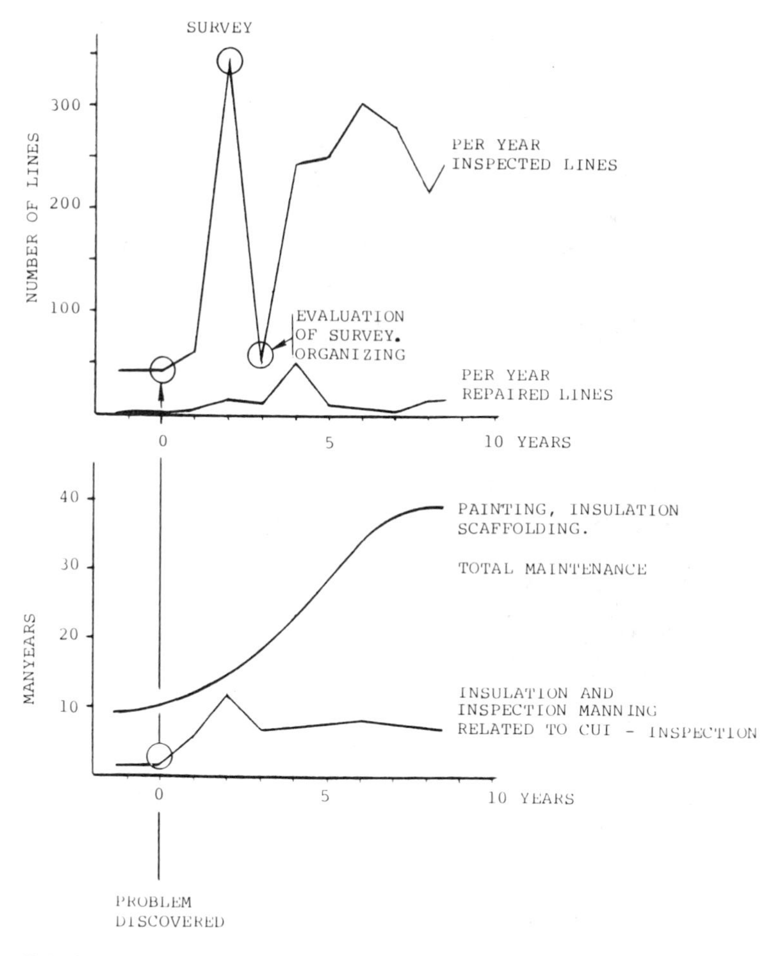

FIG. 6—*Quantity of inspection and repairs, required resources as a result of CUI.*

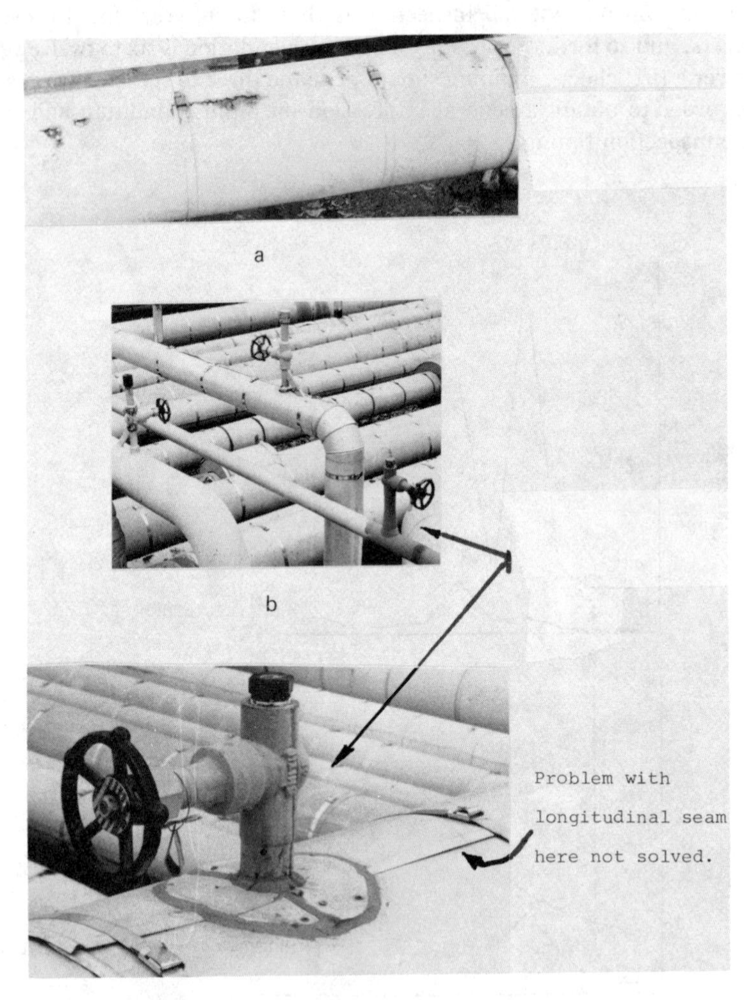

FIG. 7—*Examples of designs to prevent CUI:* (a) *painted galvanized steel jacketing in service 20 years now at the end of life, and* (b) *valve design without gusset easier to seal.*

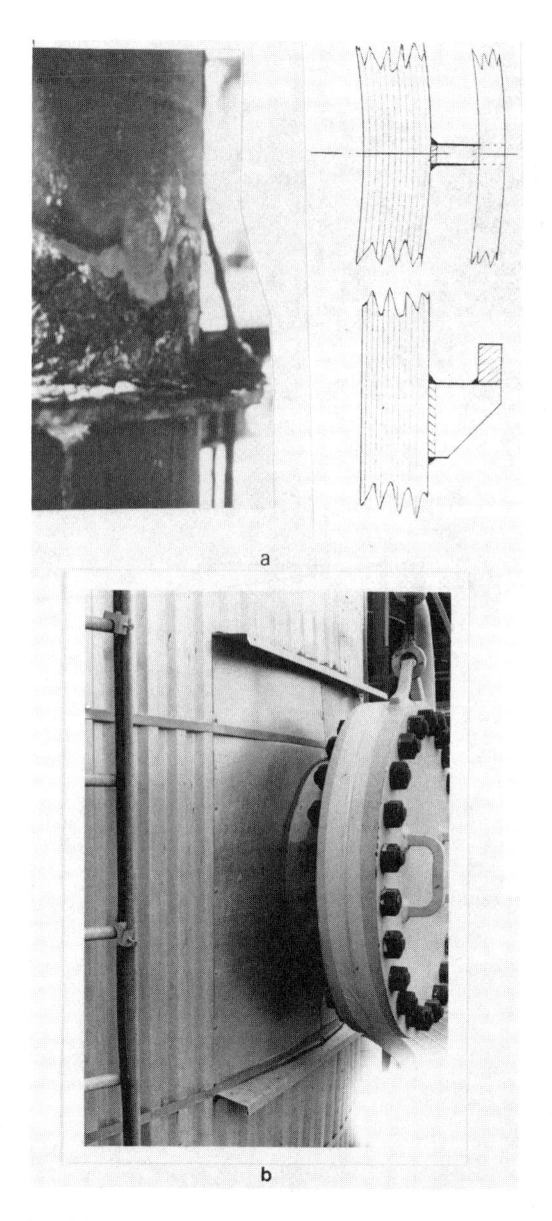

FIG. 8—*Examples of designs to prevent CUI:* (a) *this design of insulation support ring should prevent corrosion shown on the left and* (b) *jacketing details at manhole.*

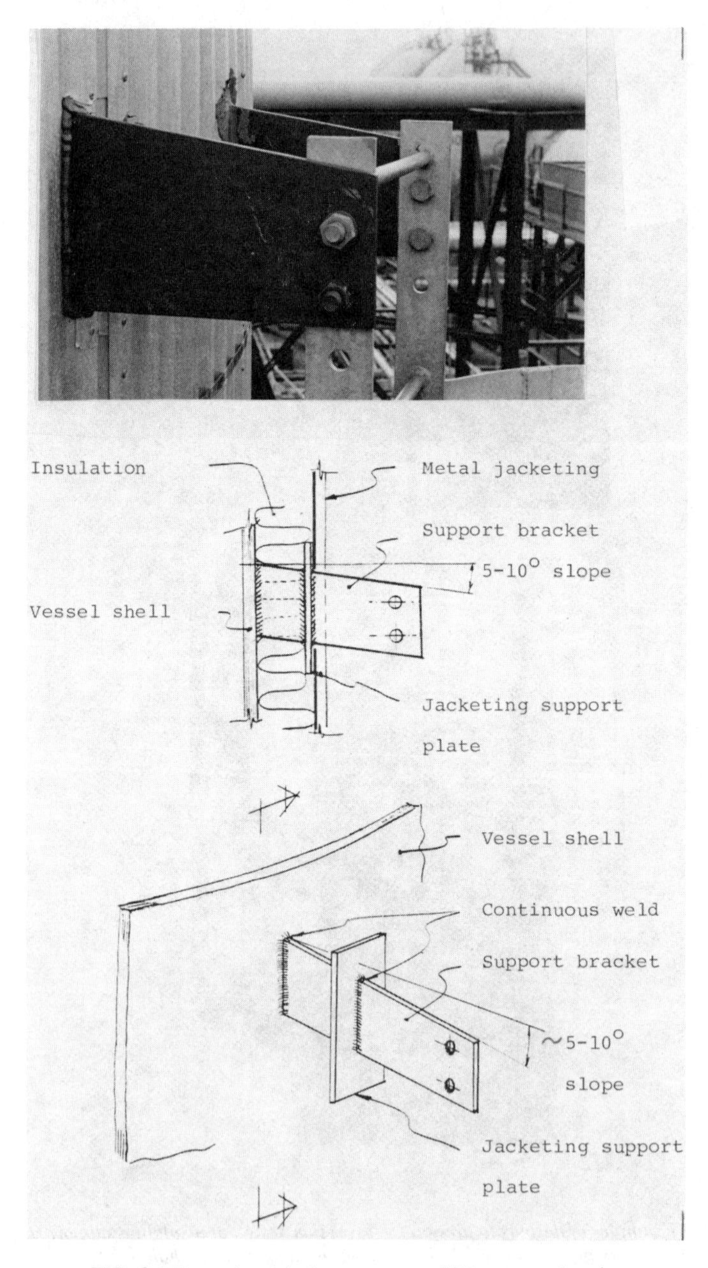

FIG. 9—*Example of design to prevent CUI: support bracket.*

Economics

During 1981, total maintenance costs for painting, insulation and scaffolding were 2 million dollars averaging 39 people working per year. This amount is four times the cost of our program for the period of 1973 through 1974. Inspection cost for corrosion control is typically 0.2 million dollars per year. The initial costs for developing programs and routines correspond to one to three man-years depending on existing routines and plant size and age. Safety and environmental aspects, together with the risk of unplanned shutdown have justified the costs of our program.

For budgeting and resource planning it is necessary to know the incidence frequency for the years after corrosion under insulation has been recognized. Our experiences at Stenungsund are probably similar to those experienced at other mature plant sites. These experiences are illustrated in Fig. 6.

This information may give guidance for those plant sites where the problem is new and some plan to tackle it is being developed. Figures 7 through 9 show some examples of improved detail design to prevent corrosion.

In the long run maintenance costs would be considerably reduced, providing improved techniques in painting, insulation and construction details were utilized both in new projects and for repairs.

Victor C. Long[1] and Paula G. Crawley[1]

Recent Experiences with Corrosion Beneath Thermal Insulation in a Chemical Plant

REFERENCE: Long, V. C. and Crawley, P. G., **"Recent Experiences with Corrosion Beneath Thermal Insulation in a Chemical Plant,"** *Corrosion of Metals Under Thermal Insulation, ASTM STP 880*, W. I. Pollock and J. M. Barnhart, Eds., American Society for Testing and Materials, Philadelphia, 1985, pp. 86–94.

ABSTRACT: Corrosion of carbon and stainless steels under wet thermal insulation can be a serious problem and can be especially prevalent in the humid Gulf Coast area. This paper discusses an inspection program that has been in progress since late 1982 at a ten-year old chemical plant located at La Porte, TX. The program is intended to determine the extent of corrosion damage to major pieces of equipment that has occurred under inhibited calcium silicate insulation finished with aluminum jacketing and to recommend remedial action.

Key elements of a successful inspection program are discussed. A series of pictures is presented that correlate visual external telltale signs with corrosion beneath the insulation. At the time of writing, significant localized corrosion of carbon steel has been found, as well as evidence of incipient stress corrosion cracking of austenitic stainless steel.

KEY WORDS: corrosion, thermal insulation, carbon steel, austenitic stainless steel, inspection

Corrosion beneath thermal insulation is a serious problem. It is insidious in nature and often goes undetected until major damage has occurred. In the case of stainless steel, total replacement of damaged equipment may be the only recourse. Costs can be high, both for repairs and lost production.

Most major chemical plants have active corrosion monitoring programs aimed at process side corrosion. Unfortunately, too little attention is often given to external corrosion under wet insulation, or more appropriately to the prevention thereof. Perhaps one reason is lack of awareness, especially at newer plant sites where there has not been enough time for corrosion related problems to develop. Cost is another reason. It is often difficult to justify ex-

[1]E. I. du Pont de Nemours Company, P.O. Box 347, La Porte, TX 77571.

by a deluge system that is tested frequently. A deluge system provides a very aggressive source of water and chlorides.

Finally, design details are important. Corrosion under insulation differs from internal corrosion in that it tends to be more localized. It is related to points of moisture intrusion and areas of moisture concentration. It is very important, for example, to know the location of insulation support rings. They tend to act as dams for moisture that has entered the system from above, and this is one area where we have seen the most indications of incipient stress corrosion cracking of stainless steel equipment. Other important design details, of course, are materials of construction, wall thickness and, in the case of stainless steel, location of stressed areas such as welds. Good characterization of equipment will help in knowing where to start and where to look.

The second phase is field inspection. We have relied heavily on visual inspection at La Porte. While access can be a problem and cosmetic repairs to the jacket system can obscure results, we have had fairly good success in correlating tell-tale external signs with corrosion beneath.

Locating points of moisture intrusion is a good first step. Cracked caulking and mastic around nozzles, manways, and on vessel heads are likely points of moisture intrusion. Mechanical damage to the aluminum jacket is another.

Staining of the aluminum jacket with leached sodium silicate and corrosion products is a good telltale sign. Also, perforation of the jacket from within caused by reaction with moisture and the caustic sodium silicate is a sure sign of trouble beneath.

Hot spots on the insulation surface are another sign of wet and damaged insulation. The human hand is a simple and sensitive instrument for this test but again access can be a problem. Temperature measurements using infrared (IR) techniques are good for remote areas, but when jacket temperatures approach ambient, this technique suffers from interferences.

Once suspect areas have been identified, core sampling of the insulation with a cork bore can be useful in confirming the presence of excessive moisture or the loss of sodium silicate inhibitor. The final proof, however, is stripping the insulation, clean up, and inspection of the metal for corrosion.

The following is a series of figures dealing with inspection of two distillation columns, one stainless steel and one carbon steel. Operating temperatures are between 80 and 90°C.

First the stainless steel column. From a distance the insulation system appeared in good shape, but closer inspection revealed several areas with telltale signs of wet insulation. About midway up the column, an insulation support ring was found protruding through the aluminum jacketing. Insulation in this area was found to be very wet, the sodium silicate inhibitor leached out, and the aluminum jacketing perforated from sodium silicate attack. Corrosion deposits were found on the stainless steel as will be seen in the next few figures.

Moisture entered the insulation system around nozzles located a few feet penditures for more effective insulation systems or an aggressive preventiv maintenance program for a problem that may not develop for 10 to 15 years. These dollars are well spent, however, because the plant integrity will be protected and energy savings will accrue from keeping the insulation dry.

I intend to share with you our experiences with corrosion under insulation at the La Porte Plant. The plant is located on the ship channel east of Houston, TX, a couple of miles from Galveston Bay. There are prevailing winds off the salt water bay, and there is ample rainfall and humidity year round. The plant is a typical chemical plant. Major equipment consists of distillation columns (some quite large), reactors, and tanks. Typical operating temperatures of the insulated equipment range from 50 to 150°C. Materials of construction in contact with insulation are generally carbon steel and American Iron and Steel Institute (AISI) 304 or 316 (Unified Numbering System [UNS] S30400 or S31600) stainless steels. The primary insulation system is calcium silicate inhibited with sodium silicate and finished with either aluminum jacketing or reinforced mastic. The plant age is about ten years.

General appearance of the insulation is good. However, on closer inspection numerous points for moisture intrusion can be found, primarily where mastic is used as finish, at seals around nozzles and manways, and where mechanical damage has occurred to the jacketing. Most of the leak points are the result of aging and the lack of a preventive maintenance program dedicated to the prevention of moisture intrusion. There are areas where the insulation has become very wet, and in some cases the sodium silicate inhibitor has been completely leached out.

Realizing that we had all of the conditions required for corrosion and because of problems at other plant sites, a program was initiated in late 1982 to address the question of corrosion under insulation. There are four key elements to this program. They are

- Equipment characterization
- Inspection
- Repair
- Preventative Maintenance

In this paper, I shall discuss the first two elements.

Equipment characterization includes operating characteristics, location within the plant, age, and mechanical design details. Thorough knowledge of your equipment is important in setting equipment inspection priorities and determining the most probable locations for corrosion for a given piece of equipment.

Equipment that operates intermittently or below about 100°C or both is generally more susceptible to corrosion under wet insulation than equipment operating continuously at higher temperatures. Location is important. Is it near a cooling tower and receives carry-over spray. Is it subject to flooding or frequent hose-downs with chloride containing process water. Is it protected above the insulation support ring as shown in Fig. 1. Note the deteriorated caulking. Figure 2 shows perforation of the aluminum jacketing and staining caused by the leached sodium silicate. The insulation was removed from the area above the insulation support ring. Crusty deposits were found just above the ring, as well as spotty iron stains on the stainless steel. In Fig. 3, the crusty deposits above the ring may be seen. They are very hard and difficult to scrape off. Iron stain areas are also noted (Figs. 4 and 5) both along the crusty deposits and more isolated specks where the stainless steel is relatively free of deposits. These iron stains are symptomatic of corrosive attack on the stainless steel. One would expect to find stress corrosion cracking in these areas, if present.

After the area was cleaned up and the metal surface polished with a sanding disk, it was dye checked. Very little if any cracking was found. Figure 6 shows several small potential cracks in the column shell adjacent to a vertical weld, a stressed area. Figure 7 shows two small pits possibly connected by a short crack in an area where there was relatively few crusty deposits. We have concluded that these were signs of incipient stress corrosion cracking. Fortunately, there is no serious equipment damage in this area, but all the conditions for stress corrosion cracking are there. We further conclude that if corrective action is not taken, serious equipment damage will occur.

Next, I plan to discuss the inspection of the carbon steel column. Again,

FIG. 1—*Deteriorated caulking, stainless steel vessel.*

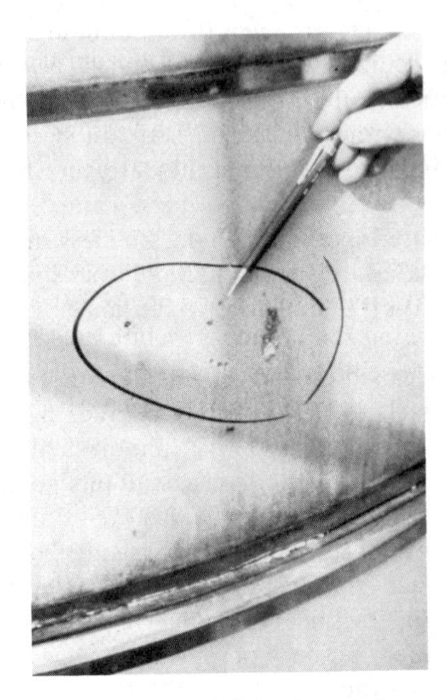

FIG. 2—*Corroded aluminum jacket.*

FIG. 3—*Crusty deposits on stainless steel vessel.*

FIG. 4—*Iron stain on stainless steel.*

FIG. 5—*Pitting of stainless steel.*

the insulation looked good from a distance, but it also suffered from moisture intrusion. Figures 8 through 11 deal with an area near the base of the column. Moisture entered the insulation system at a mastic joint where there was a slight change in column diameter (Fig. 8) and around a nozzle. Telltale perforation of the aluminum jacket had occurred (Fig. 9). After the insulation was removed, general corrosion of the carbon steel was noted. However, it was relatively mild, just a few hundred micrometres (Fig. 10) except around a manway and several nozzles (Fig. 11). In these areas heavy scale 7 mm thick extended several centimetres back underneath the insulation. Metal loss ranged from 0.5 to 2 mm. While this amount is still within the allowable thickness loss, further corrosion must be avoided or repairs will be required.

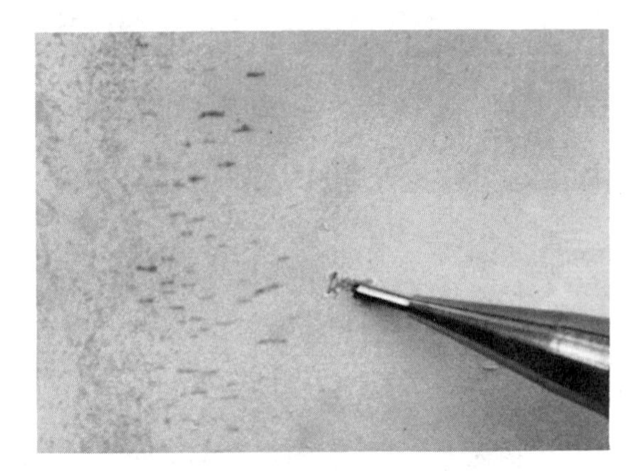

FIG. 6—*Dye check, possible cracks.*

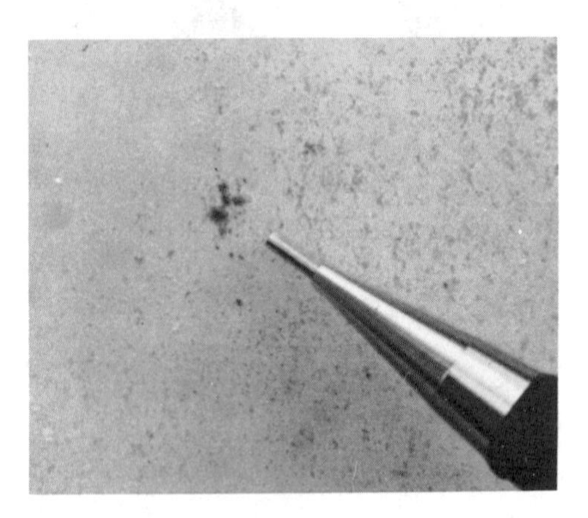

FIG. 7—*Dye check, pitted area.*

FIG. 8—*Leak point, jacket of a carbon steel vessel.*

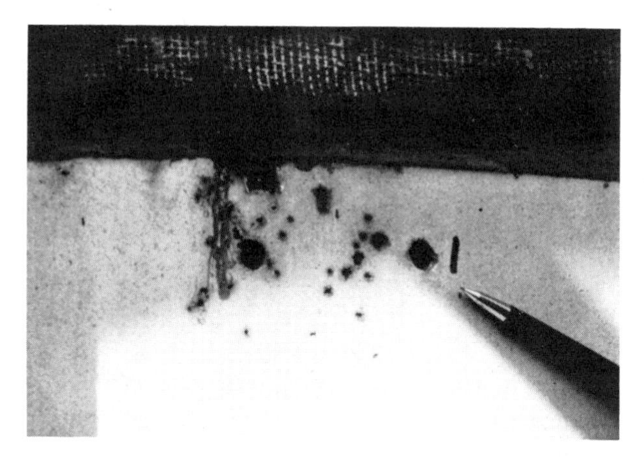

FIG. 9—*Corroded aluminum jacket.*

FIG. 10—*Minor pitting of carbon steel.*

FIG. 11—*Heavy scale, carbon steel vessel.*

Summary

We have identified numerous areas of wet insulation including some where the sodium silicate has been completely leached out. Significant corrosion of carbon steel equipment is localized, primarily to areas around manways, nozzles, and other moisture collection points. Signs of incipient stress corrosion cracking of stainless steel equipment are present, but no serious cracking has yet been found. Without prompt remedial action and a good preventive maintenance program, we conclude that significant equipment damage will occur in the future.

B. J. Moniz[1] *and M. C. Ritter*[2]

Failure of Type 316 Stainless Steel Nozzles in Contact with Fire Retardant Mastic

REFERENCE: Moniz, B. J. and Ritter, M. C., **"Failure of Type 316 Stainless Steel Nozzles in Contact with Fire Retardant Mastic,"** *Corrosion of Metals Under Thermal Insulation, ASTM STP 880*, W. I. Pollock and J. M. Barnhart, Eds., American Society for Testing and Materials, Philadelphia, 1985, pp. 95–102.

ABSTRACT: Two vertical American Iron and Steel Institute (AISI) 316L (Unified Numbering System [UNS] S31603) stainless steel (SST) nozzles on the head of a process vessel operating at 120°C leaked after six years in service. It has been plant practice to apply mastic reinforced with glass cloth as a sealer over insulation around nozzles on vessel heads. The cracking occurred where the fire retardant brominated mastic met the nozzles. Chloride stress corrosion cracking developed when the mastic seal with the nozzles broke, allowing water to enter and chloride to concentrate. The water source is the fire protection sprinkler system (50-ppm chloride), which deluged the vessel eight to ten times in six years. Up to 60 000-ppm chloride was detected on the fracture face and, significantly, no bromide. Tests showed that the mastic contained 363-ppm leachable chloride and 36 300-ppm leachable bromide. The remedy was to terminate insulation with a metal cover and seal the gap between cover and nozzles with caulking compound.

KEY WORDS: stainless steels, thermal insulation, fire retardants, calcium silicates, UNS S31600, chloride analysis, bromide analysis

Description of Failure

A 4.36-m-high by 3.65-m-diameter (14-ft 4-in. high by 12-ft diameter) 36 715-L (9700-gal) holdup tank was made of Type 316 stainless steel (Unified Numbering System [UNS] alloy S31600). It operated at 120°C and was designed for 41.3 MPa (60 psig) and full vacuum. It was situated in a location relatively sheltered from the environment at a Gulf Coast plant. After six years it began leaking process through two nozzles in the head.

[1]E. I. du Pont de Nemours and Company, Inc., Engineering Department, Beaumont, TX 77704.

[2]E. I. du Pont de Nemours and Company, Inc., Petrochemicals Department, Victoria, TX.

Figure 1 shows the location of the nozzles in the vessel head. They were 7.6 cm (3 in.) and 10.2 cm (4 in.) in diameter, both schedule 40S (wall thickness approximately 5.7 mm). After the leak was detected the vessel was shut down, the insulation stripped, and the nozzles liquid penetrant inspected to detect cracking. The two affected nozzles were cut off and examined more closely.

A second liquid penetrant inspection confirmed that the cracking had started on the outside (Figs. 2 and 3). Metallographically mounted sections taken through cracked regions indicated the highly branched morphology characteristic of chloride stress cracking (Figs. 4 and 5).

The cracking location in both cases was just below the line of contact with fire retardant mastic, which was applied as a sealant between the nozzles and thermal insulation applied to the vessel head. Figure 6 shows the location of cracking in relation to the insulation system.

Investigation of Cause

The vessel head was examined before the insulation was removed. The mastic had separated part of the way around the cracked nozzles. Since there was no elastomeric sealant between the nozzles and the mastic, it is quite

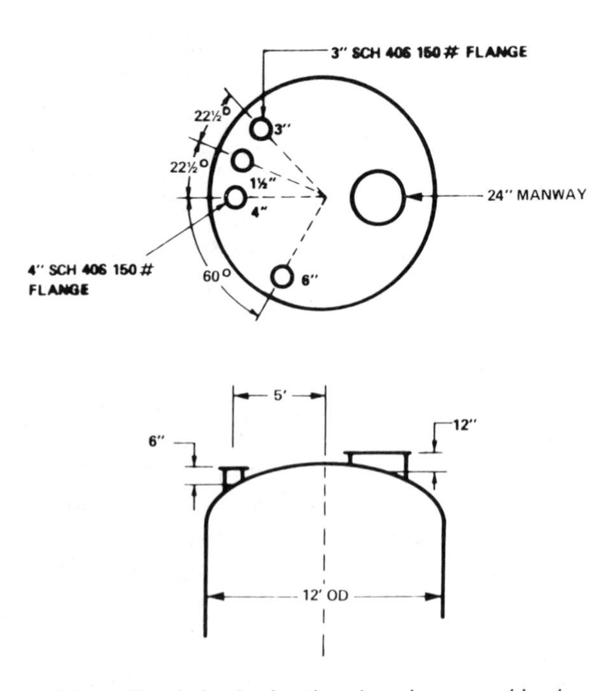

FIG. 1—*Sketch showing location of nozzles on vessel head.*

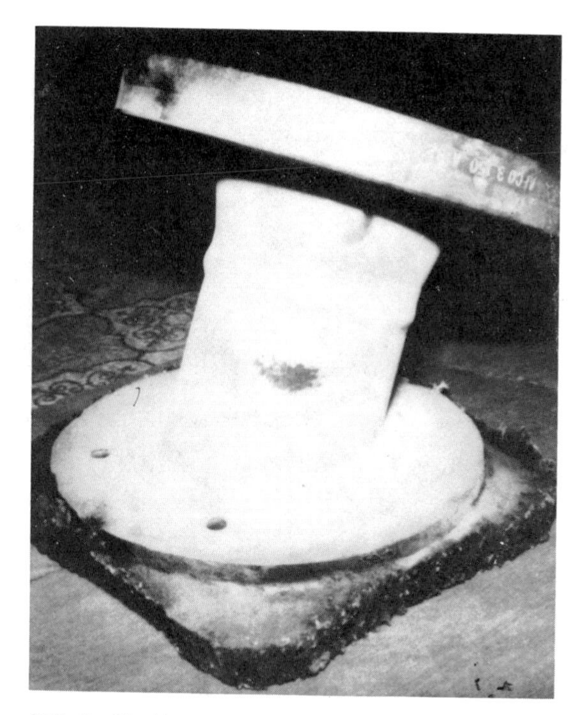

FIG. 2—*Liquid penetrant inspection, 7.6-cm (3-in.) nozzle.*

FIG. 3—*Liquid penetrant inspection, 10.2-cm (4-in.) nozzle.*

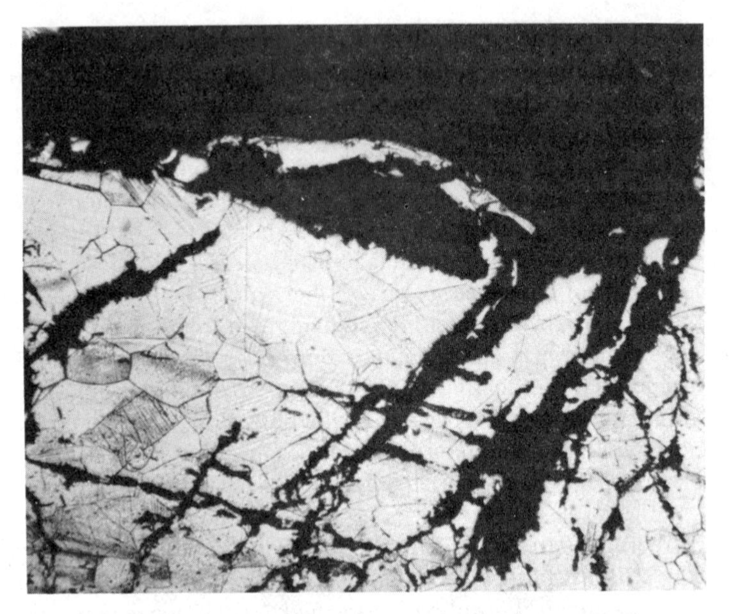

FIG. 4—*Section through 10.2-cm (4-in.) nozzle. Corrosion and stress corrosion cracking. (Approximately X100).*

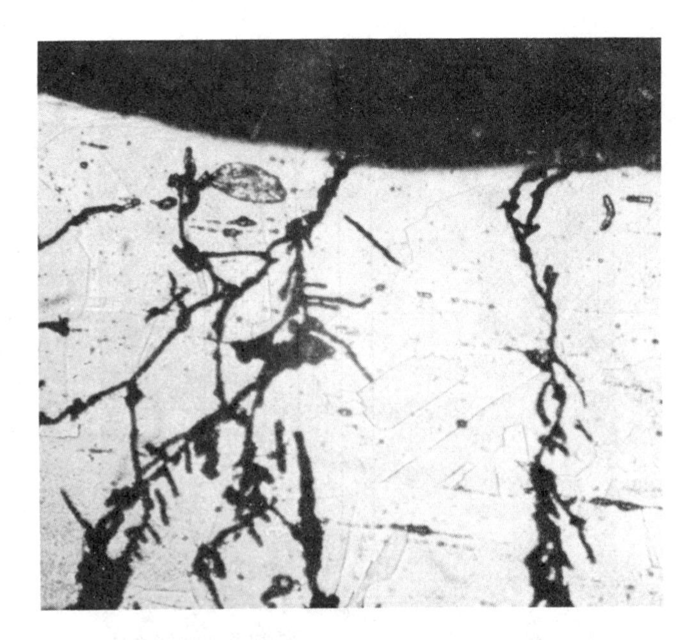

FIG. 5—*Section through 7.6-cm (3-in.) nozzle. Stress corrosion cracking. (Approximately X200).*

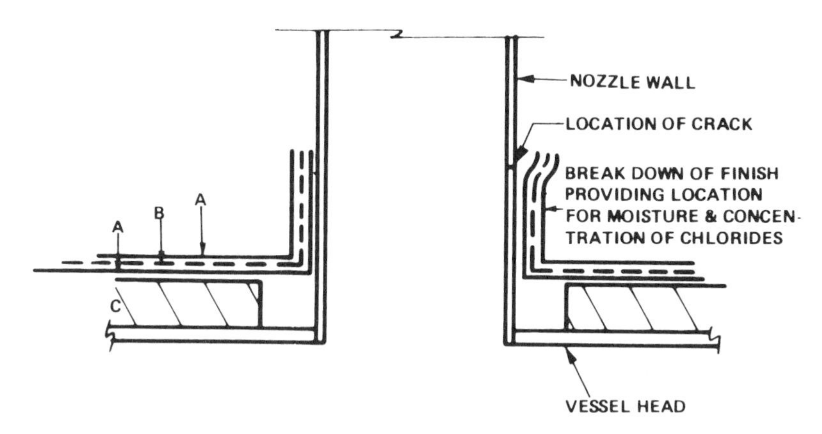

FIG. 6—*Insulation system on failed nozzle showing crack location.*

possible that the gap had existed for a large part of the time the vessel had been in service.

Samples of mastic and synthetic organic cloth (which make up the insulation termination system) were analyzed for leachable chlorides and bromides. The test is described in Table 1. The mastic contained 363-ppm chloride and 36 300-ppm bromide. The organic cloth contained 2010-ppm chloride; it was not analyzed for bromide (because it is not a fire-retardant product).

In order to detect whether bromide was partly or totally responsible for cracking, samples of the stress cracked nozzles were broken open and analyzed using energy dispersive X-ray analysis under the scanning electron microscope. There was a detection of 4.86 and 6.57 weight % chloride and, significantly, no bromide. The results are shown in Table 2.

Note that the analyses obtained by energy dispersive X-ray analysis do not include elements below atomic number 20. The results are normalized, assuming no elements below atomic number 20 exist.

TABLE 1—*Description of leachable halide test.*

5 g of solid is mixed with 200 mL water (If very low Cl^- is suspected, 100-mL water is used)
The mixture is refluxed overnight at the boiling point.
The mixture is filtered.
The filtrate is analyzed by potentiometric titration for Cl^- and Br^-.

TABLE 2—*Semiquantitative energy dispersive X-ray spectrometer analysis of fracture face.*

	Nozzle, weight %	
Element and Line	7.6 cm (3 in.)	10.2 cm (4 in.)
Silicon K alpha	3.89	2.77
Sulfur	15.3	...
Chloride	6.57	4.86
Calcium	...	1.49
Titanium	...	0.22
Chromium	12.94	32.11
Iron	61.77	47.06
Nickel	13.31	10.48
Copper	...	1.01

NOTE: no bromide detected.

Explanation of Failure

The investigation points to chloride stress cracking as the mode of failure. Despite the possible presence of bromide ions, Br^- does not appear to be responsible in any way for stress cracking.

Chloride stress cracking requires a source of water and usually a chloride ion evaporation-concentration mechanism. The source of water was most like the fire protection deluge system (Fig. 7). The system was tested approximately eight times during the six-year life of the vessel. The deluge water contains 50-ppm chloride and the vessel operates at 120°C. In addition the mastic and cloth contain 363- and 2010-ppm leachable chloride, respectively. The conditions are extremely conducive for chloride stress cracking.

Failure Prevention Program

The failure led to two courses of action:

(1) redesign of insulation termination at nozzles and
(2) analyses of all insulation materials used in the plant for halides.

Redesign has involved improving the seal between the nozzle and termination using aluminum sheathing and silicone caulk. This is shown in Fig. 8.

Halide analyses have been carried out on samples of insulation materials used at the plant site. See Table 3 for summary of the analyses. Materials with leachable chloride in excess of 250 ppm have been rejected. The presently accepted "rule of thumb" is 250 ppm in the chemical process industry. Experience suggests that material with leachable chloride below approximately 250 ppm will not cause stress cracking when in contact with 300 series stainless steel above the accepted threshold temperature of 60°C under evaporation-concentration conditions. Note that the figures described are not hard

FIG. 7—*Deluge system.*

A INSULATION MASTIC
B SYNTHETIC ORGANIC FIBER CLOTH, OPEN WEAVE
C CALCIUM SILICATE INSULATION, INHIBITED
D 16 MIL ALUMINUM SHEET

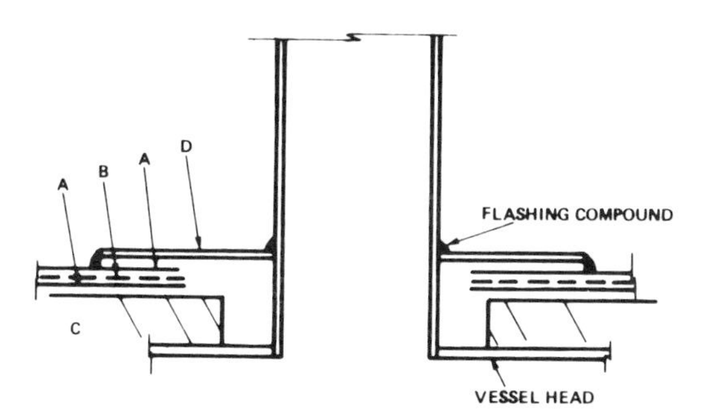

FIG. 8—*Redesign of insulation termination at nozzles.*

TABLE 3—*Test results—leachable halides in insulating materials.*

Type	Form	Leachable	
		Chloride	Bromide
Fibrated asphalt cutback "A"	cured sheet	99	<20
Fibrated asphalt cutback "B"	cured sheet	68	<20
Open weave glass cloth	as received	<34	*
Glass cloth with adhesive	as received	123	*
Fire retardant mastic "A"	cured sheet	363	36 300
Fire retardant mastic "B"	cured sheet	4 420	*
Vinyl acrylic mastic "A"	cured sheet	6 230	<20
Vinyl acrylic mastic "B"	cured sheet	340	100
Vinyl acrylic mastic "C" (nuclear grade)	cured sheet	250	94
Vapor barrier membrane	as received	<82	*
Open weave synthetic organic fiber cloth	as received	2 010	<20
Nonsetting sealer "A"	cured bead	101	*
Nonsetting sealer "B"	cured bead	<19	*
Flashing compound "A"	cured bead	111	*
Flashing compound "B"	cured bead	10	*
Flashing compound "C"	cured bead	68	*
Loaded mastic polymer based	cured sheet	171	<20
Paint, silicone based	cured film	120	<25
Paint, polyamid catalyzed epoxy	cured film	190	<25
Paint, silicone rubber	cured film	<2	<25
Paint, epoxy polyamid	cured film	10	*
Wax paper used to cure sheet specimens	as received	<226	*

*Not determined.

and fast but are given as the guidelines of the industry. In some cases (for example, good company-wide experience) materials having leachable chloride in excess of 250 ppm are not rejected.

Lately, company guidelines have further evolved to paint stainless steel with two coats of polyamide catalyzed epoxy paint, minimum dry film thickness 0.08 to 0.11 mm (3 to 4 mil) up to a service temperature of 121°C (250°F). Also, fiberglass reinforced polyester or thixotropic two-part epoxy reinforced open weave cloth has superceded aluminum jacketing as the recommendation for heads. These materials can be buttered up to nozzles and reportedly do not break away like glass cloth reinforced mastic.

William G. Ashbaugh[1]

External Stress Corrosion Cracking of Stainless Steel Under Thermal Insulation—20 Years Later

REFERENCE: Ashbaugh, W. G., "**External Stress Corrosion Cracking of Stainless Steel Under Thermal Insulation—20 Years Later,**" *Corrosion of Metals Under Thermal Insulation, ASTM STP 880*, W. I. Pollock and J. M. Barnhart, Eds., American Society for Testing and Materials, Philadelphia, 1985, pp. 103–113.

ABSTRACT: In the mid-1960s, a rash of costly stress corrosion cracking failures of stainless steel under insulation occurred. A series of tests were performed to establish the failure mechanism and to provide information for prevention. Since that time, the Union Carbide Corporation (UCC) Central Engineering policy has been to use protective coatings on the stainless steel. No special insulation materials, inhibitors, or protective weather barriers have been used to prevent external stress corrosion cracking (ESCC). During these past 20 years, there has been only a few incidents of ESCC of painted stainless steel; therefore a satisfactory pay back on the cost of painting the stainless steel has been realized. There are still questions to be answered, and optimization of the painting practice is taking place. Although painting is not the only way to prevent ESCC, it is a proven procedure that can be accomplished with today's technology.

KEY WORDS: austenitic stainless steel, corrosion, stress corrosion, insulation, thermal insulation, chloride cracking, external stress corrosion cracking, external stress cracking, protective coatings, paintings, cracks, concentration cell corrosion

During the past 20 years, we have had considerable experience in dealing with attempts to prevent external stress corrosion cracking (ESCC) of stainless steel under thermal insulation. With the occasion of this meeting we thought that it would be appropriate to summarize these experiences and information in the hope that it will assist industry in combating this costly and wasteful form of corrosion attack. Several years ago, a survey was made to evaluate the annual cost of corrosion. These dollar figures are huge. The point that I found interesting was that it was estimated that as much as three quarters of this cost can be prevented using known technology. I believe that in this conference there is suf-

[1]Manager of corrosion and materials engineering department, Union Carbide Corporation, P.O. Box 471, Texas City, TX 77590.

ficient technology to essentially eliminate ESCC, if the knowledge is applied and used properly. We will begin this paper with a review of our early problems and then take you through our testing program, the conclusion we have reached and the procedures and criteria we have employed successfully for the last 20 years to reduce and nearly eliminate this problem.

The Problem

In the early 1960s, a large chemical production unit was built at one of Union Carbide's Gulf Coast plants. Much of the process equipment in this unit was American Iron and Steel Institute (AISI) Type 316 L (Unified Numbering System [UNS] 5316) stainless steel. In less than six months, several of the stainless steel distillation columns developed leaks in their shells. As we investigated this, we were shocked to find large areas of tiny cracks visible on the outside of the column sections. The cracks appeared very much like the root system of small plants (Figs. 1 and 2). The cracking resulted in drop-wise seeping of process liquid through the cracks, which ultimately leaked from under the insulation calling attention to the leaks. What first seemed to be very tiny leaks soon turned out to be significant areas of badly transgranularly stress corrosion cracked stainless steel.

As more and more insulation was removed from adjacent vessels, piping, and columns, other areas of ESCC were found, in many cases, just the beginning traces that had not yet penetrated the vessel walls. Once all the areas where stress corrosion cracking had occurred were identified and repaired or replaced, we turned to the question of what caused the cracking. In all cases, the cracking occurred under cellular glass insulation with the equipment operating at moderately elevated temperatures. We knew we were looking for chlorides; did they come from the Gulf Coast air, from nearby process units during

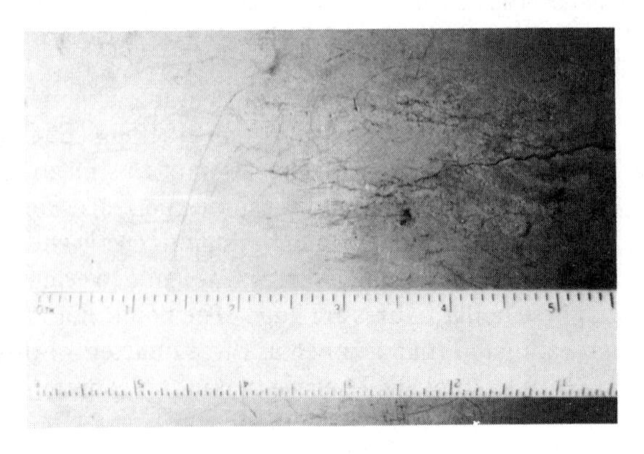

FIG. 1—*External stress corrosion cracking of a Type 316 vessels under cellular glass and anti-abrasion coating.*

FIG. 2—*Another example of ESCC. Here the sample had been bent to open up the main crack.*

construction and before insulation was applied, from contact with the ground before erection, or from the insulation? What was the source of the chlorides?

A review of all possible sources included the cellular glass insulation and the anti-abrasion coating on the inside surface. When we tested the anti-abrasion coating, we found that it contained polyvinyl chloride. Upon warming it up in the presence of water, we found hydrochloric acid! Other sources of chlorides, while feasible, were not believed to be highly involved since there was nothing unusual about the exposure of the stainless steel equipment during its fabrication or erection compared to many other stainless steel items.

The evidence all pointed to the warm, stainless steel equipment operating in contact with the anti-abrasion coating. Rainwater inside the insulation was the third necessary ingredient. Any anti-abrasion coating based on polyvinyl chloride was summarily banned from use in our insulation activities from then on.

The magnitude and the cost of this particular ESCC problem focused our attention on the mechanism of cracking stainless steel under insulation and brought a number of questions to mind. What other insulation materials might cause the stress corrosion cracking, under what conditions, and what must be done to prevent a recurrence of this incident?

The Test Program

At this same time, other companies were suffering from similar ESCC problems, and were attributing it to one type of insulation or another. These reports, together with our experience, indicated that we needed to determine which insulation materials were safe and which would actively contribute to stress corrosion cracking.

A simple laboratory test was developed using a horseshoe bend stainless steel specimen that was heated and then placed in contact with the insulation material (Fig. 3). Distilled water was fed to the interface between the insulation ma-

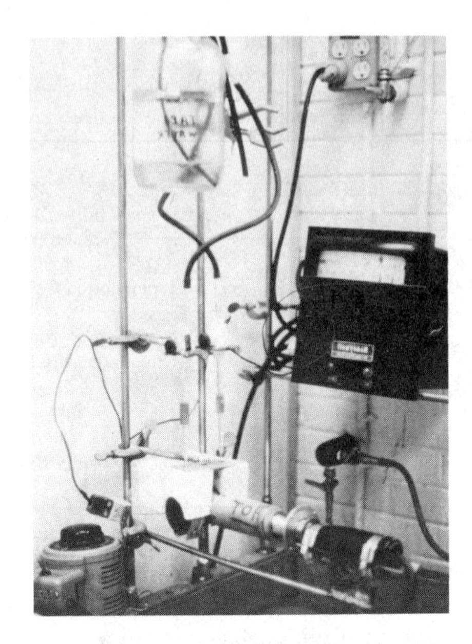

FIG. 3—*Lab test apparatus used to determine if insulation materials would cause stress corrosion cracking of Type 316 stainless steel.*

terial and the stainless steel. After an appropriate exposure time and temperature, observations were made as to whether or not cracking had developed. This test procedure was later reported in an article in *Materials Protection* [*1*].

Evaluation of some 36 insulation materials using this procedure showed that only four would, in themselves, cause stress corrosion cracking. If this were true, or even close to being correct, it indicated to us that insulation as a class was not responsible for the majority of failures.

While we were engaged in the laboratory testing phase, we had been reviewing our company history of stainless steel corrosion problems and found scattered, but widespread reports of stress corrosion cracking that were either correctly diagnosed as ESCC or were described as simply stainless steel cracking failures without a metallurgical examination. We observed that many of these cases of ESCC occurred in the absence of thermal insulation. For example, one of the most prevalent trouble spots was underneath steel slip-on flanges at stainless steel lap joints (Figs. 4 and 5).

As we completed the test program and reviewed plant operating history, it became apparent that if we were to attack the ESCC problem by concentrating on the type or quality of the insulation material, we would be overlooking the bigger culprit, that is, the total surrounding presence of chlorides from a variety of sources.

Reconstructing the mechanism of attack, we suggest that the chloride and water somehow must enter into the annular space between the insulation and

FIG. 4—*Example of ESCC of a 316 pipe at lap joint. Chloride and water collected under slip-on flange causing stress corrosion cracking.*

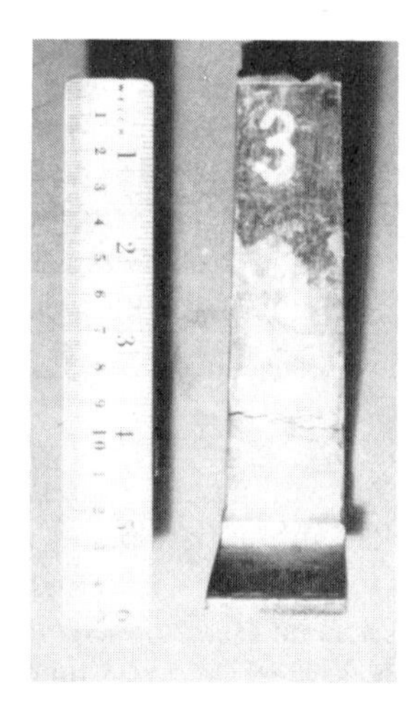

FIG. 5—*A stainless steel column section sample showing ESCC that occurred under a lap ring flange. This area had not been painted as specified.*

the outer surface of the stainless steel equipment. In addition, heat and stress must also be present to allow the stress corrosion cracking to initiate and proceed. We have further concluded that heat plays a large role in the evaporation of water and subsequent concentration of chlorides to the point where they become quite aggressive. Since we have no control over the austenitic stainless steels normal stress patterns, the only other parts of the equation we can deal with are the presence of chlorides and water.

The use of "inhibited" insulation materials was investigated and rejected because our lab tests showed they would not prevent stress corrosion cracking when chlorides were introducted under the insulation. Use of inhibited insulation would limit the insulation designers freedom of materials selection if it were required for all stainless steel equipment. And finally, it would not offer protection under flanges, and so forth. This then left us with the consideration of waterproof insulation systems to exclude water and chlorides or applying a protective coating to the stainless steel.

To achieve a watertight protective coating for an insulation system, one would have to produce, in effect, a pressure tight covering. This then would have to be maintained in this pressure tight condition to prevent ingress of water and chlorides. Our insulation advisors were quick to point out that such protective covering would be exorbitantly expensive and that maintaining such a system, if it were installed, would probably be impractical.

We then concluded that a protective coating on the stainless steel under the insulation would be the most workable approach. The protective coating system also would deal with the stress corrosion cracking problems under flanges, pipe clamps, and other external appurtenances.

Thus, we came to the conclusion in the middle 1960s that the way to protect stainless steel from ESCC was to paint it. The mere thought of painting stainless steel was at first repelling and strange to many people who had not experienced the problems we have already described. The selection of which paint for stainless steel also become quite a point of discussion, that is still going on today. However, at that time we selected two paints that we felt would be suitable based on previous experience with them in other applications.

Painting

In order to effectively proceed with painting stainless steel to protect it from ESCC, it was necessary to select appropriate paints and generate the painting criteria or guidelines for what to paint and where to paint. The selection of protective coatings was based on several criteria such as ease of application, good adhesion to mill finished stainless steel, and resistance to hot water vapors.

The ability to maintain coating integrity in the presence of hot water vapors and to withstand temperatures up to about 200°C was important. The 300°C, we felt, was the safe upper limit above which stress corrosion cracking would not occur because the equipment and insulation would be dry. If equipment were cycling from higher to lower temperatures, it would have to be dealt with as a special case.

Our first coating selection was a modified, carbon filled silicone that cured at ambient temperature (Fig. 6). This product was tested on our stress corrosion cracking apparatus and proved to be effective in preventing cracking from chloride bearing water. One weakness of the modified silicone was its poor resistance to chemical fumes and spills. We therefore added an epoxy phenolic coating for those areas where we felt chemical exposure might occur. The selec-

FIG. 6—*A portion of the first process unit to have all its stainless steel painted—both insulated and uninsulated areas were painted.*

tion of protective coatings turned out to be much easier than the development of a criteria for when and where to use them.

Our original project criteria, dealing with this subject, stated that the responsible materials engineer assigned to the project or the plant corrosion engineer would identify those critical pieces of equipment that needed to be painted. Our intention was to spend the money for the protective coating only where it would be most needed and where the pay off in preventing stress corrosion cracking would be most obvious. Typical items to be painted were stainless steel pressure vessels, equipment handling corrosive or toxic materials where leaks would not be acceptable, or one of a kind pieces of equipment whose outage would seriously affect a major production unit. Any number of discussions were held about various projects and plant maintenance jobs to sort out what was critical and not critical, as well as what was or was not worth painting. At times, it seemed that it would be cheaper to go ahead and paint rather than to discuss it; but we were making a real effort to be selective in the use of the protective coatings through a conscious and logical decision process.

As a result, a regular patch work of painted stainless steel developed in the late 1960s, and we found that inevitably we had some failures of items that were not painted, which we wished we had painted. We also found that painting this item and not painting that item, painting this pipe and not painting another pipe in the same project was confusing and the cause of mistakes in the construction phase. In 1970, a new unit was being designed and a decision was made at that time to paint all the stainless steel (Fig. 7). Since most of the unit would be stainless steel, this was a major decision and change in criteria. This change was rather like running the first 4-min mile; once we overcame the barrier and reached understanding and acceptance of the new philosophy, paint-

FIG. 7—*A view of a Type 316 L column that had been in service about 12 years. It was painted with the modified silicone.*

ing of all stainless steel, as a general rule, became easier. Exceptions could now be made as to where paint was not required, rather than as previously, when painting was the exception.

Experience

Up to this point we have been describing our past experience, history, and philosophy. Where are we today? After almost exactly 20 years of using paints to protect stainless steel from stress corrosion cracking, our experience has been excellent. We are convinced that a proper paint film protects stainless steel for many years and that the small cost for this kind of insurance has been paid for many times over.

But all is not perfect in this world and we have had some problems. We have had piping and column sections painted where the painters did not pull back the lap flange rings and paint underneath them: the result was ESCC. Painted vessels had the paint damaged by steel cable slings, tearing and scarring the paint: the result was ESCC. A series of small tanks that were insulated with cellular glass, and contrary to explicit instructions, a polyvinyl chloride containing anti-abrasion coating was used as a cement to glue the cellular glass to the painted vessel. The solvents in the adhesive/anti-abrasion coating penetrated and destroyed the protective coating allowing the soon to be formed hydrogen chloride to do its stress corrosion cracking work: the result was ESCC. A substitution was made for paint specified, the paint failed: the result was ESCC. Painting was specified for small process piping; the project manager needed to save money and cut out the painting: the result was ESCC.

Another problem area is the total paint job on stainless steel equipment, which has steel flanges and attachments. We know that the one coat of modified silicone will not offer protection to carbon steel (Figs. 7 and 8). The epoxy phenolic must be used to protect the steel. (Note: We do not allow zinc primers on steel in contact with stainless steel.) Thus, a careful reading of our engineering

FIG. 8—*A closer view of the column shows how this silicone paint has disappeared from the steel reinforcing rings. The paint on the stainless steel is still in good condition.*

criteria would have portions of a stainless steel vessel painted with the epoxy phenolic and other portions painted with modified silicone (Figs. 9 and 10).

This makes for a colorful paint job but presents possibilities for confusion by the fabricator or painter and probably extra cost compared to a single paint system. As a result, we now use the epoxy phenolic for the entire vessel provided it is compatible with the process temperature (Figs. 11 and 12).

Summary

In summary, our position and experience regarding stress corrosion cracking of stainless steel under insulation indicates that the most practical way to

FIG. 9—*A new heat exchanger that has its stainless steel surface painted with modified silicone and its steel painted with epoxy phenolic.*

FIG. 10—*A combination of steel and stainless steel, each with its own paint specification.*

FIG. 11—*A short column section of Type 316 L with all parts painted with epoxy phenolic.*

FIG. 12—*A large distillation column, Type 304 stainless steel being insulated. Both stainless steel and steel portions are painted with epoxy phenolic.*

protect the stainless steel is with a coat of good paint. While our results over the past 20 years have not been perfect, the few and isolated failures have essentially been physical problems, which actually served to demonstrate that while the paint film is intact, stress corrosion cracking is prevented. It is also worth noting that the limited cracking that occurs at defects in the paint film is repairable. A general stress corrosion cracking failure often resists repair thus requiring replacement.

The use of a protective coating on stainless steel to ward off chloride stress corrosion cracking may be looked at like the wearing of a seat belt in an automobile. The seat belt does not guarantee that somebody will not come up and hit you, but if it happens, your risk of serious injury and death is considerably less. So the protective coating on stainless steel has to be evaluated against the risk and consequences of ultimate stress corrosion cracking failaure. We cannot predict when or how chlorides might invade the insulation system, but we can trust the coating to protect the stainless steel when the inevitable happens.

Reference

[1] Ashbaugh, W. G., "ESCC of Stainless Steel Under Thermal Insulation," *Materials Protection*, May 1965, pp. 19-23.

Donald O. Taylor[1] *and Rodney D. Bennett*[1]

Shell and Jacket Corrosion of a Foamed In-Place Thermally Insulated Liquefied Petroleum Gas Tank

REFERENCE: Taylor, D. O. and Bennett, R. D., **"Shell and Jacket Corrosion of a Foamed In-Place Thermally Insulated Liquefied Petroleum Gas Tank,"** *Corrosion of Metals Under Thermal Insulation, ASTM STP 880*, W. I. Pollock and J. M. Barnhart, Eds., American Society for Testing and Materials, Philadelphia, 1985, pp. 114–120.

ABSTRACT: An 87.78 m (288 ft) diameter by 24.77 m (81.25 ft) high liquid butane storage tank was insulated in 1976 using vertically foamed in-place (FIP) low flame spread polyurethane with an aluminum outer jacket. The tank had not been painted since the vendor stated the urethane would not properly bond to painted surfaces. In 1979 corrosion of the aluminum jacket and aluminum rivets was noted. Chemicals, leached from the urethane foam, resulted in streaks of discoloration down the sides of the tank. Detailed inspection in 1980 involved cutting out sections of jacket and insulation to uncover the tank shell. Extensive pitting was noted on the steel shell as well as the foam side of the aluminum jacket. Corrosion is attributed to the intrusion of water from a seawater deluge system that leached the acidic fire retardant chemicals from the urethane foam. A task force was organized to investigate and evaluate various methods for insulating large cyrogenic liquid storage tanks. The study involved visitations both in the United States and Europe. As a result, detailed specifications were developed for the blasting, painting, installation of foam glass insulation, and weather proofing of the tank. Since corrosion had apparently reduced the tank wall thickness below that required by American Petroleum Institute (API) Recommended Rules for Design and Construction of Large, Welded Low Pressure Storage Tanks (API 620), detailed analyses of mechanical properties of the steel were investigated, a program was established to acoustically monitor the tank during hydrotest, and procedures were established for minor localized weld repair to the deepest pits. Details of the inspection procedure, testing, painting, and insulation specifications are discussed.

KEY WORDS: insulation, tanks (containers), corrosion, polyurethane, butane storage

[1]Engineering consultant group leader of Process Department and supervisor of Technical Support Unit, Technical Engineering, respectively, Arabian American Oil Co., Process Department, E-6100 Engineering Building, Dhahran, Saudi Arabia.

Experience with corrosion under insulation is widely varied. In carbon or low alloy steel tanks the corrosion may take the form of simple rusting, or dependent upon the material that infiltrates behind the insulation, severe pitting and even stress cracking have been encountered. Stainless steel tanks have also failed because of leaching of chlorides from insulation or ingress of other chloride containing contaminants or both. Corrosion of the metallic moisture barrier or deterioration of a nonmetallic moisture barrier permits ingress of water and airborne chemicals that accelerate attack.

This paper will cover the details of the original insulation of a liquid butane storage tank, discovery of progressive corrosion, tank testing, and the reinsulation procedures and specifications.

Discussion

In 1976 an 87.78 m (288 ft) diameter by 24.77 m (81.25 ft) high liquid butane storage tank operating at $-40°C$ was insulated using vertically foamed in-place polyurethane with fire retardant additives and an aluminum sheeting weather barrier (Figs. 1 and 2). The carbon steel tank was not painted before insulation since the vendor stated the urethane would not properly bond to painted surfaces. No provision was made to prevent moisture ingress other than a sealant between the overlap of the aluminum sheeting, which was pop riveted together.

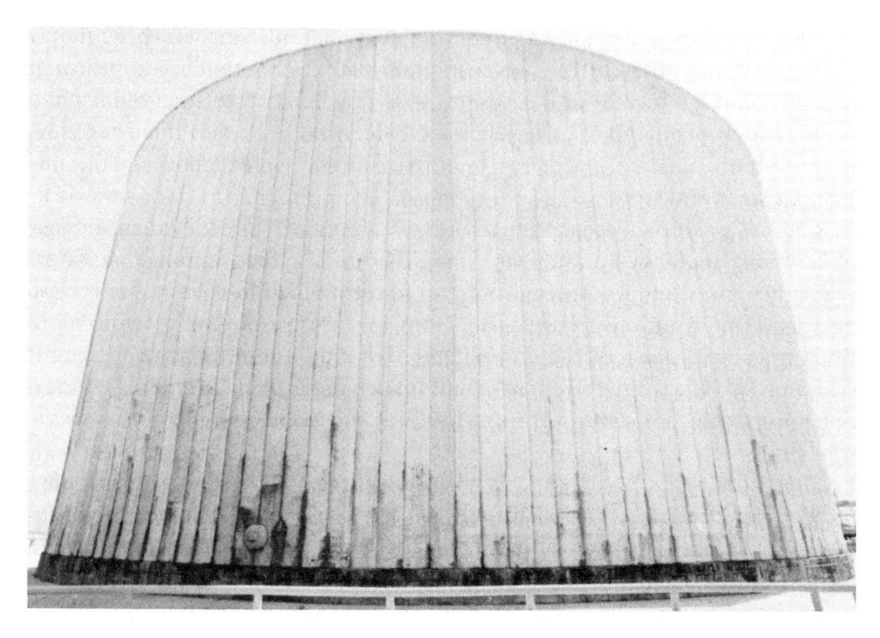

FIG. 1—*Rust breakthrough on vertical foamed in-place urethane insulation.*

FIG. 2—*Close-up of Fig. 1.*

In 1979 there was external evidence in the form of large rust stains indicating failure of the insulation system. Random areas of the sheeting and polyurethane were removed to permit visual inspection and ultrasonic thickness measurement of the steel shell. The polyurethane foam was generally found to be water saturated for at least 152.4 mm (6 in.) on either side of the aluminum overlap joint. Some areas in the lower region of the tank were wet throughout.

Earlier data gathered from inspection of insulated butane lines indicated expected corrosion rates of 0.254 mm (0.01 in.) per year although lines had been painted. Thus plant experience indicated that corrosion rates in excess of 0.254 mm (0.01 in.) per year might reasonably be expected on the unpainted steel tank shell.

Randomly sampled areas were found to have a general metal corrosion loss of 2.54 mm (0.100 in.) plus as much as 2.54 mm of pitting in some locations!

Further investigation verified that the water leached the acidic fire retardants (fluorine-bromine compounds) from the urethane. These acidic additives, plus breakdown of the urethane foam by intense sunlight and periodic seawater deluge, resulted in acid solutions having a pH as low as 1.7. Additional high chloride water was introduced by the weekly testing of the seawater deluge system.

Coinciding with the early stages of this investigation there was an on-going project to reinsulate 80 000 butane liquid (bbl) tanks with a horizontal foamed in-place urethane system (Fig. 3). The procedure was modified to paint the tank shell with red oxide primer before insulation.

An additional skirt modification to shed water was added to the bottom of the tanks to avoid the collection of water at the interface of the aluminum

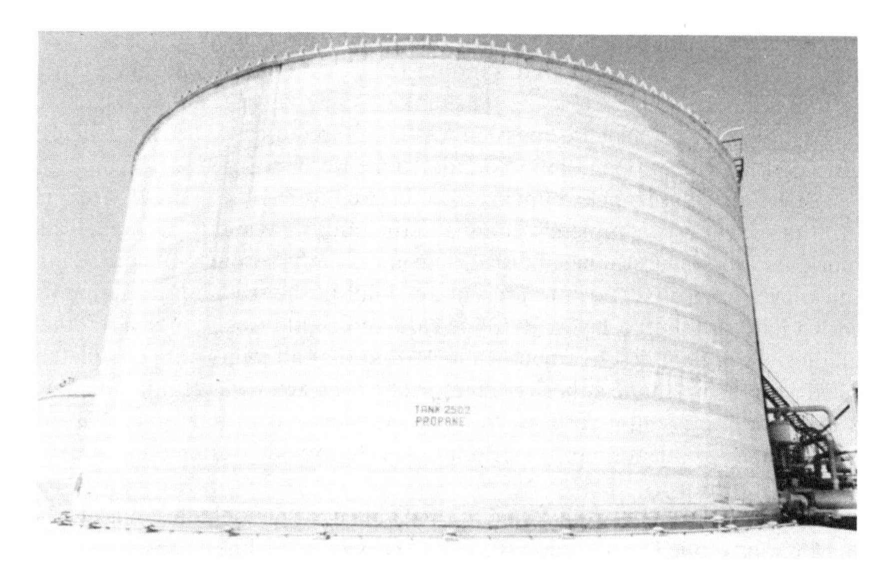

FIG. 3—*Rust breakthrough on horizontal foamed in-place urethane insulation.*

sheeting and the ring wall foundations. This skirt modification was used in the construction of 7 new 750 000 bbl tanks. Five additional terminal tanks were constructed in 1980 through 1981 with an epoxy primer, full ring wall skirt and the horizontal foamed in-place urethane system.

In mid 1980 an Engineering Task Force Group was formed to conduct a worldwide investigation of cold insulation systems. After consulting with more than 20 different companies and visiting some 22 sites with cold service storage facilities operating at temperatures from −10° to −2°C (14 to 28°F), the task force concluded that no two companies used the same system. Although a decision had not been reached as to the optimum method of insulation, it became critical to remove one of the 895 000 bbl tanks from service because of evidence of accelerating corrosion rates on the steel shell. The tank was stripped of insulation, sandblasted, marked off by square foot, and had ultrasonic thickness measurements made and recorded. All pitted areas were also minutely inspected and gaged. Results showed general corrosion loss over the entire tank. The areas in contact with the wet polyurethane insulation, more than 40% of the tank, showed heavy metal loss of 2.54 to 5.08 mm with extensive deep pits, a corrosion rate of 0.68 to 1.35 mm per year (based on actual service life).

In late 1980 the decision was made to use 152.4 mm (6 in.) of adhered cellular glass insulation covered with a 2.03 mm (.080 in.) thick moisture barrier of polyester mesh reinforced acrylic/aliphatic urethane. Detailed stress calculations indicated that the steel shell could adequately be repaired by fill-ing the critically deep pits with weld metal that was then ground flush. Accordingly, the shell was repaired, sandblasted to white metal, and coated with an inorganic zinc silicate (Fig. 4). The tank was then hydrotested to the maximum fill level, and acoustic emission was monitored during the test to detect any crack or flaw propagation. No indications were recorded.

The cellular glass blocks were installed in two layers. The first 101.4 mm (4 in.) thick layer was completely bonded to the tank shell and to each adjacent block using a polyurethane adhesive. The joints in the second 50.8 mm (2 in.) thick layer were offset 50% to prevent any joint in the second layer lining up with a joint in the first layer. Extreme care and rigid inspection insured that all blocks were bonded to adjacent blocks and that all joints were completely filled with the urethane adhesive (Fig. 5). The contractor was required to have a technical representative from the insulation manufacturer present at all times during application of the cellular glass to provide training and inspection assistance.

After all cellular glass block had been installed, a coat of high solids acrylic mastic was applied at 0.76-mm (0.030-in.) dry film thickness, and while still wet a knitted open mesh polyester cloth was rolled in to achieve complete wetout and a smooth surface. A second coat of acrylic mastic was then applied to achieve a total dry film thickness, including fabric, of 1.78 mm (0.070 in.).

After complete curing, inspection, and repair of any imperfections or bubbles, two coats of an aliphatic urethane were applied to a dry film thickness of

FIG. 4—*Scaffolding—before application of foam glass.*

FIG. 5—*Application of foam glass, note "buttered joints."*

0.254 mm (0.010 in.). Thus, the total thickness of the completed vapor barrier system was 2.03 mm (0.080 in.). This system is not only highly impermeable to moisture but is also very resistant to damage by wind blown objects. In case of such damage it is readily repaired, access to the damaged area being the biggest problem.

A very important aspect of the job, which is often neglected, is the proper sealing of areas where there are penetrations through the insulation. Such might be ladder supports, gage connections, nozzles, and so forth. The sealant must be readily applied (preferably by a gun), must have excellent bond to the surfaces to which it is applied, must be nonshrinkable (thus must have minimal solvents), must have excellent weatherability, and should be capable of accepting a paint cover coat if desired (Fig. 6).

Summary

The most significant factor in the massive corrosion of the shell and jacket was determined to be the acidic fire retardant solution leached from the polyurethane foam by the moisture penetration of the inadequately sealed jacket

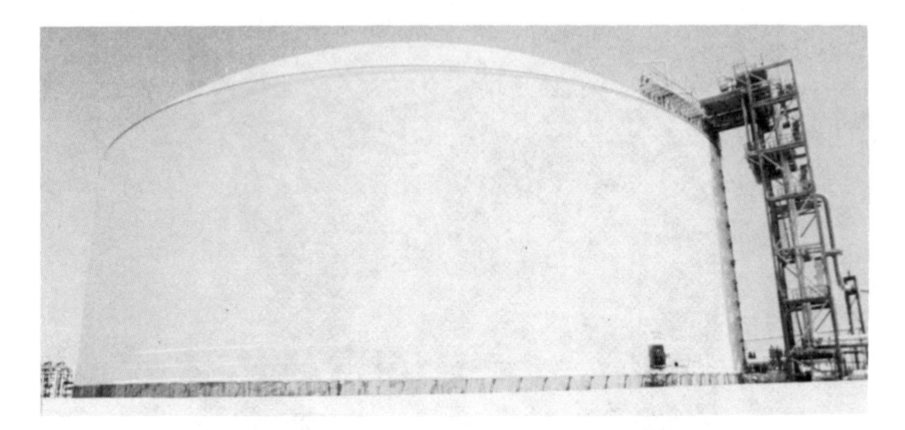

FIG. 6—*Completed foam glass-barrier coated tank.*

joints. The high humidity of the marine environment and the repeated exercise of the deluge system accelerated the water ingress. The decision not to paint the tank before application of the polyurethane foam permitted direct exposure of the steel to the corrosive environment.

The new system has now been in service for over one year with no signs of any deterioration. The fiber reinforced polyurethane exterior retains its gloss finish and pleasing appearance. There is no evidence of moisture infiltration at any point.

Thus we feel it has been effectively demonstrated that a proper, cost effective, insulating system can be accomplished if necessary care is given to system selection, specification, training, and inspection plus time and money.

Acknowledgment

The authors gratefully acknowledge the cooperation of their fellow colleagues for their assistance in gathering data and photographs. We also wish to thank the Arabian American Oil Company of Saudi Arabia for their permission to publish and present this paper.

William G. Ashbaugh[1] *and Thomas F. Laundrie*[1]

A Study of Corrosion of Steel Under a Variety of Thermal Insulation Materials

REFERENCE: Ashbaugh, W. G. and Laundrie, T. F., **"A Study of Corrosion of Steel Under a Variety of Thermal Insulation Materials,"** *Corrosion of Metals Under Thermal Insulation, ASTM STP 880*, W. I. Pollock and J. M. Barnhart, Eds., American Society for Testing and Materials, Philadelphia, 1985, pp. 121–131.

ABSTRACT: Corrosion of steel equipment under thermal insulation is and has been a costly problem in Union Carbide Corporation (UCC) chemical operations as well as other refining and chemical plants throughout the world. The Materials Technology Institute (MTI) has spent about $30 thousand trying to develop a nondestructive inspection method to detect and measure corrosion under insulation. The corrosion rate of carbon steel under insulation depends upon a number of variables: moisture, oxygen availability, metal temperature, paint system, and the insulation. Little work has been published that defines the effects different types of insulation have on the corrosion rate of carbon steel. Some suggest that the type of insulation does make a difference. An in plant test was designed and operated to obtain quantitative data comparing several types of insulation and their effect on corrosion of steel.

Twelve pieces of insulation were applied to carbon steel pipe and were exposed to the atmosphere at Texas City, TX. Water, in addition to rain, was occasionally added to the outside of the pipe, and steam was run through the inside of the pipe once a week in order to accelerate the corrosion rate.

This paper will report qualitatively and quantitatively on the corrosion rates of the carbon steel under the various types of insulation. The data will be useful to insulation specialists in guiding their selection of the kind of insulation from the standpoint of its influence on steel corrosion.

KEY WORDS: corrosion, thermal insulation, steels, paints, coatings, pipes, thermal cycling, industrial atmosphere

Corrosion of steel equipment under thermal insulation is and has been a costly problem in the Union Carbide Corporation (UCC) chemical operations as well as other refining and chemical plants throughout the world. The cor-

[1]Manager of corrosion and materials engineering, and corrosion and material engineer, respectively, Union Carbide Corporation, P.O. Box 471, Texas City, TX 77590.

rosion rate of carbon steel under insulation depends upon a number of variables: moisture, oxygen availability, metal temperature, type of paint system, and the insulation. Until now, little information has been published on the effects different types of insulation have on the corrosion rate of carbon steel. An accelerated test was run at a Gulf Coast plant using twelve different types of insulation for a one year period. Qualitative and quantitative measurements were made on the corrosion rates of carbon steel under the various types of insulation.

In general, the types of insulation used had a small effect on the corrosion rates of carbon steel compared to the other factors such as moisture and oxygen availability. The corrosion rate was worse where the moisture was allowed to collect on the pipe, such as underneath where the insulation hung away from the pipe or at the edges of the test pieces. Some of the insulation types drew the water away from the pipe and reduced the corrosion rate. An extended test of three years or more would be needed to bring out the subtle differences that may exist between the types of insulation.

A protective coating of paint is effective in preventing corrosion, regardless of the type of insulation and is still our recommended procedure.

Background

The problem of corrosion under thermal insulation is well known in the refining and petrochemical industry. It has been a growing concern as more and more cases come to light. Two recent examples are shown in Figs. 1 and 2. An increasing number of papers are being written on the subject. The Materials Technology Institute (MTI) has spent about $30 thousand trying to

FIG. 1—*Example of corrosion under insulation. Gas heater periodically steamed out, only had an inorganic zinc primer originally, and was insulated for personnel protection after it had been in service for awhile. The hole resulted in a fire.*

FIG. 2—*Example of corrosion under insulation. This area is in the kettle section of a column, which was slightly above ambient. Upper portions of the column run well below ambient and condense moisture, which runs down the wall. The proper coating had been used, but it had been applied too thin. Pinholes allowed the corrosion to get started.*

develop a nondestructive method to detect and measure corrosion under insulation, so far without success [*1,2*]. A company memo based on UCC's experience and test work was written in 1980 by W. G. Ashbaugh entitled "Corrosion of Steel Under Thermal Insulation, The Hidden Danger."

Introduction

It has been suggested that the corrosion rate of carbon steel is influenced by different types of insulation. Virtually no quantitative data exist on this subject. This accelerated test was designed to obtain such data under carefully controlled conditions. This data will be useful to insulation specialists in guiding their selection of the kind of insulation from the standpoint of its influence on steel corrosion.

Procedure

The Test

Twelve pieces of insulation, each approximately 0.6 m long (2 ft), were applied to 7.62-cm (3-in.) carbon steel pipe. The insulation was wrapped with an aluminum weather barrier, but the ends were left open. The insulation samples were spaced approximately 0.3 m (1 ft) apart on the pipe. The materials were then exposed to the atmosphere at the plant paint test site, located just north of a cooling tower, for approximately one year (Figs. 3 through 5).

Before the insulation was installed, the steel pipe was sandblasted to a near-white metal finish and painted with an epoxy-phenolic paint covering about one-fourth of the outer circumference as a longitudinal stripe.

FIG. 3—*Insulation test site north of the cooling tower. Test stand before steam tubing was attached.*

FIG. 4—*Closer view of Fig. 3.*

Seventy-pound steam was run through the pipe for 15 min once a week in order to expose the steel to a cycle of cold/wet to hot/dry in the expectation that this would disturb the semi-protective oxide films that develop as steel rusts. When the steam was turned on, the pipe wall temperature would level out at 93°C within 5 min.

Originally, it was hoped that rainwater would be an adequate source of moisture for the corrosion of the pipe. However, after three months, a few pieces of insulation were removed, and it was discovered that very little corro-

FIG. 5—*Insulation being applied to the pipe by skilled insulators.*

sion was occurring under the insulation. Rain fell during this period only 14 of the 90 days.

A number of holes were then punched through the weather barrier and the insulation, and 0.24 L of water was squirted into the holes of each piece of insulation twice a week, if it did not rain. The water used was the plant untreated river water.

The above steps, particularly the water addition and thermal cycling were taken to provide an accelerated corrosion test so that useful data could be obtained in one year's time.

Moisture sensors[2] were installed under the insulation on some of the test pieces to detect the presence or absence of water between the pipe and the inner surface of the insulation (Fig. 6).

The Evaluation

The wall thickness of the pipe was measured before and after the test with an ultrasonic thickness instrument by the Plant Inspection Department. Visual observations were made, and data were recorded of the heating and watering cycles of the pipe during the test. Observations were recorded for each piece of insulation concerning the appearance of the pipe immediately after the removal of the insulation at the end of the test (Fig. 7). Thickness measurements were taken under the middle and near the edges of each of the insulation samples. Some of the worst looking areas near the edges of the insulation were measured in more detail (Fig. 8). It was also noted how far from the edge of the insulation the severe corrosion occurred. The general

[2]Sereda Miniature Sensor, Model SMMS-01, Epitek Electronics Inc., Ogdensburg, NY.

FIG. 6—*Setup near the end of the test. Plastic cups protecting moisture sensor leads can be seen.*

FIG. 7—*The pipe and baked expanded Perlite insulation at the end of the test.*

degree of pitting was recorded under the middle of the insulation piece on the top of the pipe, the bottom of the pipe, and the side of the pipe. A forced ranking of the insulation was made.

Discussion and Results

The corrosion of the bare pipe between the insulation pieces was more severe than that underneath any of the insulation samples. The rate over the

FIG. 8—*Part of the pipe after the test and cleanup. Ultrasonic thickness measurements were taken at this time.*

one year period averaged about 50 mil per year (mpy, where 1 mil = 0.001 in. = 0.0254 mm). The iron oxide formation seemed to be rather rapid in the first half of the year and then slowed down somewhat.

A steel corrosion coupon was exposed in the test area to determine the corrosion rate on bare steel by weight measurements. This coupon showed a corrosion rate of 18 mpy as opposed to the 50 mpy on the pipe. This difference was probably caused by two factors:

- The reference sample was hanging in a vertical position, allowing it to dry faster than the pipe.
- The sample was not subject to the heating cycle that the pipe was put through every week to deliberately increase the corrosion rate.

The corrosion rate of the reference coupon is much higher than the rural atmospheric corrosion rate on steel; but is typical of a chemical industrial/marine atmosphere, particularly downwind of a cooling tower. This spot was chosen for the accelerated test.

The heating cycle was selected to dry out the surface of the pipe and its rust scale, yet not dry out all of the moisture that had accumulated in the insulation. Observations during the test confirmed this mode of operation.

Each week, if it did not rain, some untreated Brazos River water was squirted among the holes of each of the insulation samples. The pipe was watered approximately twice a week, whether it was from rain or artificial watering. It was later discovered that the corrosion rate directly under the water holes was virtually the same as that on the bare steel. Two points specifically measured showed a corrosion rate of 47 and 53 mpy. These spots were not included in judging the effect the insulation had on the corrosion rate of the steel.

Several moisture sensors (Fig. 9) were used as a "Go-No-Go" indication of

FIG. 9—*Moisture sensor used in the test. Alternating strips of gold and copper set up a potential difference when an electrolyte is present (that is, water).*

moisture under the insulation. They showed that the pipe surface did dry out rapidly while the steam heated up the pipe. The sensors failed over a period of three months. However, enough data were obtained to show that the holes and water addition did help bring moisture to the pipe-insulation interface, whereas before the holes were punched, it had stayed relatively dry.

A paint stripe covering approximately one-quarter of the pipe's outside surface was applied to the entire length of the pipe. It was one coat of Plasite 7122,[3] an epoxy phenolic. We normally specify that two coats of an epoxy phenolic be put on carbon steel under insulation. The one coat system was used to evaluate the performance of the thinner coating. This paint stripe was 6 to 7 mil thick and held up very well, both under the insulation and between the pieces of insulation on the exposed pipe. There was some undercutting along the edges, but this was to be expected. Overall, the paint performed well and the corrosion rate under it was nil.

The corrosion under the edge of the insulation was almost as severe as that outside the insulation. This was expected because there was nearly equal access to water and oxygen at this point. Moving from the exposed steel to the middle of a piece of insulation, the corrosion rate generally dropped off quickly. The distance that the high corrosion rate continued to occur from the edge of the insulation on inward, in most cases, was 2.5 to 5.0 cm (1 to 2 in.).

In all of the types of insulation tested, the corrosion was usually worse along the bottom of the pipe than on the sides or top. Along the bottom where a piece of insulation would hang away from the pipe, the moisture would form a drip line and not be drawn away by the insulation. The pipe was installed on slight angle (approximately 5°, see Fig. 4) to allow for draining of the condensate when steam was put into the pipe. This also affected the direction the rainwater entered, and the corrosion under the edge of the insulation was generally worse on the "high end" than on the "low end" of each piece.

See Table 1 for a summary of the results. Table 2 lists the detailed data

[3]Wisconsin Protective Coating Company, Union Carbide Corp. Paint Specification PCM-214.

acquired for the evaluations. The following observations listed in Table 2 were used in decreasing order of importance to establish the ranking in Table 1: (a) typical metal loss; (b) extent of the pitting on top, sides, and bottom; (c) extent of the heavy corrosion in from the edge of the sample; and (d) worst metal loss at the edge of the sample.

Two types of expanded perlite insulation were tested. For each type, we tested one as received and one after it had been baked for about 24 h at 260°C to simulate what might happen to the insulation on a steam line or a heated line after a period of time. The corrosion rate under the unbaked perlite insulation dropped off quickly less than 2.5 cm from the edge, but the baked perlite, which was falling apart, allowed high rates to proceed up to 7.5 cm under the insulation. Some of the pitting that occurred in this area was as bad as on the bare steel pipe. The baked insulation was partially broken apart and mushy in texture. The corrosion rates in the middle of the baked and unbaked perlite samples, however, were not significantly different from each other.

The expanded Perlite gave the best performance of the insulation types. It produced the lowest corrosion rates where the insulation touched the pipe near the middle of the sample. The four different types of calcium silicate performed about the same. The worst performers were the nonabsorbent cellular glass and the mineral wool whose loose fibers stuck into the top iron oxide layers.

TABLE 1—*Results.*

Rank	Insulation (No.)	Comments
1	expanded perlite (1)	Line of pitting 2.5 to 5.0 cm wide on bottom
2	expanded perlite (2)	Line of pitting 2.5 to 5.0 cm wide on bottom
3	calcium silicate (4)	Rank items 3, 4, 5, and 6 are about the same in performance. Loose fitting in that insulation hung away from bottom. Drops of water hanging on pipe underneath. Insulation very heavy with water when removed.
4	calcium silicate (3)	
5	calcium silicate (2)	
6	calcium silicate (1)	
7	fibrous glass	More pitting along bottom and edges than calcium silicates.
8	expanded perlite baked (1)	Insulation fell apart easily. Much more pitting near the edges. Some more pitting underneath.
9	expanded perlite baked (2)	Insulation fell apart easily. Much more pitting near the edges. Some more pitting underneath.
10	polyurethane foam	Only one where pitting was fairly uniform all around the pipe. Very little advanced corrosion under the edges.
11	cellular glass	Corrosion in middle about as bad as that near the edges.
12	mineral wool	Some insulation stuck to pipe.

TABLE 2—*Corrosion under insulation data and comments.*

	Amount of Corrosion									
	Typical Metal Loss, in.[b]		Worst Metal Loss at Edges, in.		Extent of Edge Effect, in.	Pit Depth in Middle[c] 1-Least 5-Most			Ranking by Appearance[d] Before Sandblasting	
Rank[a]	Middle	Edges	Up	Down		Top	Side	Bottom		Comments
1	0.008	0.010	0.046	0.010	1	2	2	5	1	unbaked expanded perlite line of pitting along bottom 2.5 to 5.0 cm wide.
2	0.018	0.029	0.032	0.029	1	2	3	5	4	unbaked expanded perlite line of pitting along bottom 2.5 to 5.0 cm wide
3	0.016	0.018	0.027	0.028	1.5	2	2	4	2	calcium silicate types all about the same
4	0.013	0.025	0.031	0.042	2	1	2	3	3	calcium silicate types all about the same
5	0.017	0.020	0.027	0.031	2	2	3	3	5	calcium silicate types all about the same
6	0.021	0.026	0.025	0.026	1	2	3	3	6	calcium silicate types all about the same
7	0.014	0.034	0.054	0.030	3	2	3	4	7	worst case of high rates in from edge for unbroken insulation
8	0.009	0.026	0.050	0.028	2	2	3	4	9	baked expanded perlite lot of edge effect and broken insulation
9	0.016	0.037	0.062	0.028	3	1	2	4	8	same as No. 8. rank
10	0.019	0.028	0.024	0.023	1	3	3	3	10	uniform pitting on all sides, very little edge effect
11	0.026	0.033	0.053	0.021	...	2	5	5	11	corrosion in middle about as bad as at edges
12	0.035	0.040	0.053	0.055	1 to 2	2	4	4	12	some insulation stuck to pipe

[a] See Table 1 results for insulation identification.
[b] 1 in. = 2.54 cm.
[c] Visual after sandblasting.
[d] Visual after removal of insulation.

Conclusions

Although this test produced no final answers, the results have added to our understanding of the several mechanisms that are taking place between the steel-water-insulation combination. We may conclude the following from the test results:

- Corrosion rates differed slightly under the various types of insulation. A similar test of three years or more would be needed to bring out the subtle differences between materials.
- The more corrosion occurred the longer moisture was in contact with the pipe. This was mainly on the underside of the pipe and at the elevated end of the insulation sample.
- Insulation that trapped the moisture against the pipe, without drawing it away by absorbing it, produced the most corrosion. The pipe in contact with insulation that was absorbent had a relatively lower corrosion rate.
- The corrosion rate of the bare steel between insulation test pieces was higher than normal because of the thermal cycling.
- The epoxy-phenolic paint did an excellent job of protecting the steel and preventing corrosion.

Summary

With a good weather barrier, the nonabsorbent insulation may actually help keep the moisture away from the pipe. Where there are breaks in the insulation covering, water can get under the insulation and be trapped against the steel. The corrosion rate is then largely dependent on how long the moisture is in contact with the steel and the thermal operating conditions. Absorbent insulation types aid in removing the water from the metal surface. They lose their advantage, however, if there is a continuous source of water such as condensation or process leaks.

The present UCC Engineering policy of painting steel under insulation (PCM-500, *UCC Painting and Coating Technology Manual*) continues to be the most effective way to prevent corrosion.

Acknowledgment

We wish to acknowledge the assistance of E. C. Powell, Jr. in planning the test program and J. H. Rapp in obtaining the insulation materials and having them installed.

References

[1] "Investigation of Nondestructive Testing Techniques for Detecting Corrosion of Steel Under Insulation," MIT Report 4, July 1981.

[2] "Investigation of an Approach to Detection of Corrosion Under Insulation," MIT Report 7, March 1982.

James R. Myers[1] *and Arthur Cohen*[2]

Behavior of a Copper Water Tube Exposed to Natural Carbonaceous Granular and Cellulosic Insulation Materials

REFERENCE: Myers, J. R. and Cohen, A., **"Behavior of a Copper Water Tube Exposed to Natural Carbonaceous Granular and Cellulosic Insulation Materials,"** *Corrosion of Metals Under Thermal Insulation, ASTM STP 880*, W. I. Pollock and J. M. Barnhart, Eds., American Society for Testing and Materials, Philadelphia, 1985, pp. 132–142.

ABSTRACT: Copper is essentially immune to corrosion in most underground environments because of the protective films that naturally form on the metal's surface. In general, there is no major concern regarding the corrosion of copper underground unless it is exposed to certain aggressive substances. These include moist environments containing (1) appreciable quantities of chlorides, sulfates, sulfides, or ammonium compounds or both; (2) organic or inorganic acids or both; or (3) active anaerobic bacteria or all three. It is also known that copper should not be directly embedded in cinders that generally contain sulfides or carbon or both. Fortunately, when corrosive conditions exist, a variety of cost effective techniques can be used to mitigate unacceptable corrosion of buried copper.

In spite of these well known general guidelines regarding the underground behavior of copper, situations do occur where the metal is placed in contact with aggressive thermal insulations. When these insulations are hygroscopic, have low resistivities and acidic pH values, and contain sulfur compounds along with carbon, it is understandable that unacceptable corrosion of the copper will occur.

The conditions that caused natural carbonaceous granular "thermal insulation" to deteriorate the underground copper tube domestic hot water system for a low-rise high density housing development are presented. Basically, the copper water tubes corroded because the "thermal insulation" absorbed large quantities of moisture, had a low resistivity (less than 1200 Ω·cm), and an acidic pH (as low as 4.8). Corrosion was facilitated by the presence of appreciable quantities of sulfur (8000 ppm) and possibly carbon in the "insulation."

Corrosion of the adversely affected copper water tube system was mitigated by cathodic protection. In order to minimize shielding effects created by the complexity of the underground system, deep anode beds were incorporated into the cathodic protection design.

[1]President, JRM Associates, 4198 Merlyn Dr., Franklin, OH 45005.

[2]Manager of standards and safety engineering, Copper Development Association Inc., Greenwich Office Park 2, Greenwich, CT 06836-1840.

The unaffected underground copper tube domestic cold water and steel gas distribution systems were bonded to the cathodically protected copper hot water tubes in order to prevent the possibility of stray current corrosion.

In addition, other insulation systems have been shown to be deleterious to copper. For example, certain chemically treated cellulosic insulation materials can contain constituents that become aggressive to copper when either moist or under high relative humidity conditions.

KEY WORDS: corrosion, insulation, copper, thermal insulation

The outside surface of copper water tube is essentially immune to corrosion in most building system environments because of the protective films that naturally form on the metal surface. In general, there is no major concern regarding the external corrosion of copper water tube in building systems unless it is exposed to certain aggressive conditions or environments. These include moist environments containing (1) appreciable quantities of chlorides, sulfates, sulfides, or ammonium compounds; (2) organic or inorganic acids; or (3) active anaerobic bacteria [*1–4*]. It is also known that copper should not be directly embedded in cinders that generally contain sulfides or carbon or both.

In spite of these well established guidelines regarding copper, situations occasionally occur where the metal is in contact with an aggressive thermal insulating environment. For example, if the insulation is hygroscopic, has a low resistivity, has an acidic pH value when wet, and contains sulfur along with carbon, unacceptable corrosion of the copper can occur.

This paper identifies two such adverse conditions involving thermal insulating substances that have been known to cause the corrosion of copper water tube. Certain natural carbonaceous granular materials used for underground thermal insulation contain appreciable quantities of carbon and sulfur. These insulations have low resistivities and acidic pH values when moist. Such an environment will be aggressive and should be avoided.

Certain chemically treated cellulosic-type insulations are aggressive to copper if they contain specific constituents (for example, ammonium sulfate) and become wet or are exposed to a high humidity environment. The corrosive conditions created by a wet ammoniacal environment are further aggravated if the insulation/moisture mixture has a low pH. Water extracts of certain cellulosic type insulations have pH values as low as 3.7.

Moisture is a necessary prerequisite for even a potentially aggressive thermal insulation to be corrosive. It is often difficult to prevent the ingress of water as in the case of underground thermal insulations especially if they are installed below the water table. Chilled water lines can condense moisture in high humidity environments unless the thermal insulations are covered with an effective vapor barrier.

When reasonable care is taken in the selection and installation of thermal insulations, however, copper tube corrosion can be prevented.

Natural Carbonaceous Granular Insulation

Natural carbonaceous granular material (that is, finely pulverized subbituminous coal) has been used for insulating underground copper water tube. Typically, properly compacted, finely pulverized, subbituminous coal should provide copper with excellent thermal insulating characteristics if it remains dry. Corrosion would not be expected to occur in a 10^{13} $\Omega \cdot$cm resistivity environment.

Examination of the analysis of subbituminous coal, however, suggests that it could be aggressive to copper if the pulverized product became wet (Table 1). The wet material would be expected to corrode copper because it contains appreciable quantities of carbon and sulfur. Further, the leaching of soluble components from pulverized subbituminous coal would be expected to create a low pH, low resistivity environment. Premature underground corrosion of thermally insulated copper water tube at two locations investigated by the authors can be associated with these aggressive conditions.

Domestic Hot Water Lines, Brooklyn, NY

Copper water tube was selected and installed in 1976 to convey domestic hot water at a seven block, low-rise high density housing development in Brooklyn, NY. A finely pulverized, natural carbonaceous granular material was specified to provide thermal insulation for the underground lines. Within three years, numerous leaks occurred in the hot water system. The leaks occurred where the copper water tube was underground and thermally insulated.

Examination of a tube that had been removed from service revealed significant corrosion on the outside surface (Figs. 1 and 2). There was no evidence of any significant corrosion of the inside surface by the water conveyed.

In general, the corroded outside tube surfaces were covered with a black product that was overlain with a greenish material (Fig. 2). Energy dispersive X-ray analysis (EDXA) and microchemical analysis revealed that the greenish tinted copper corrosion product contained major quantities of copper and sulfur (as sulfate) along with constituents normally found in soil (Fig. 3). The green copper corrosion product consisted primarily of copper sulfate. Similarly, analysis (Fig. 4) revealed that the black copper corrosion product consisted primarily of copper sulfide.

Subsequent on-site examination revealed that wet thermal insulation (nearly water saturated) existed at those sites where major corrosion had taken place. Tube to soil potentials at these locations were generally more positive than -0.15 V, referenced to a copper-copper sulfate electrode (that is, the copper exhibited potentials at which active corrosion would be expected to be taking place) [7]. On-site testing also revealed that thermogalvanic and stray current corrosion were not involved in the corrosion pro-

TABLE 1—*Typical ultimate analysis for subbituminous coal [5,6].*

Constituent	Weight, %
carbon	64 to 76
hydrogen	4.4 to 5
nitrogen	1.2 to 1.5
sulfur	0.4 to 1
oxygen	12.8 to 17

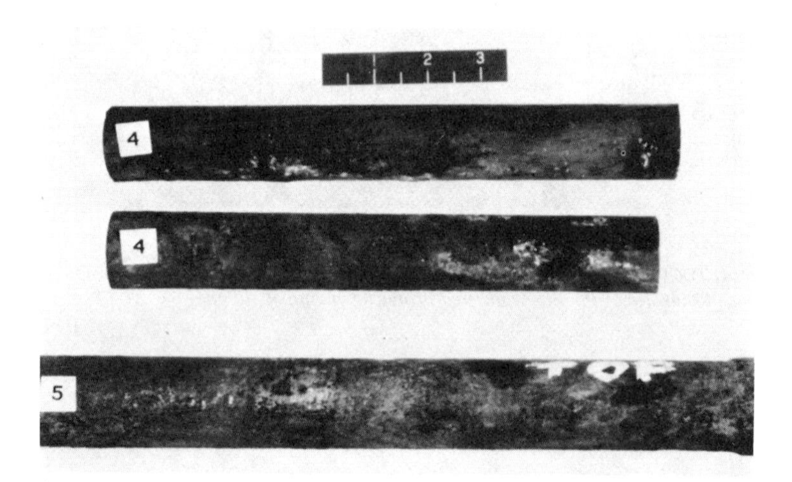

FIG. 1—*External surfaces of corroded underground copper water tube that had been thermally insulated with a natural carbonaceous granular material. The performation in Specimen 5 (see encircled area) had initiated on the outside surface of the tube (~X0.4).*

FIG. 2—*External surfaces of additional corroded underground copper water tube that had been thermally insulated with a natural carbonaceous granular material. Typically, the corrosion products consisted of black copper sulfide overlain with green copper sulfate (~X0.4).*

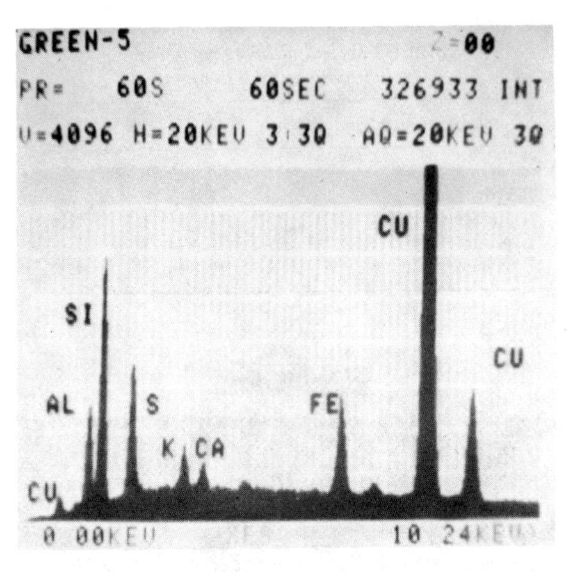

FIG. 3—*EDXA and microchemical analyses revealed that the green copper corrosion product on the outside surface of the tube consisted primarily of copper sulfate.*

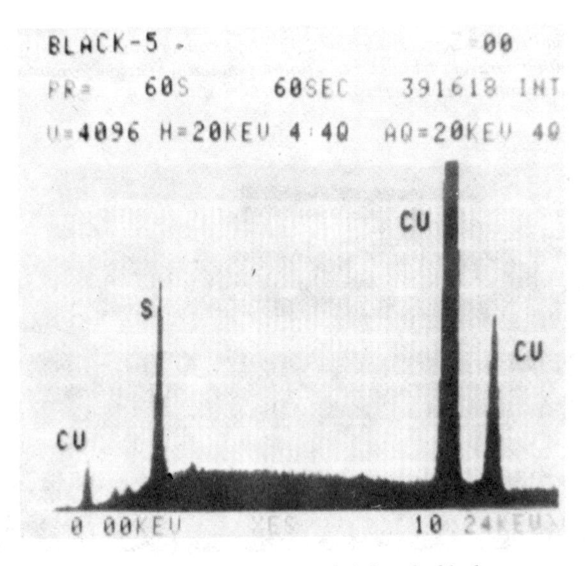

FIG. 4—*EDXA and microchemical analyses revealed that the black copper corrosion product on the outside surface of the tube consisted primarily of copper sulfide.*

cess. Analysis of soil and insulation specimens collected during the on-site inspection provided additional insight regarding this corrosion incident.

Typically, the resistivity of the wet thermal insulation ranged from 1 200 to 5 100 Ω·cm. By contrast, wet soil resistivity adjacent to the insulation varied from 9 500 to 12 000 Ω·cm. These results established that the low resistivity of the thermal insulation was not associated with the leaching of minerals from the soil but could only be related to the inherent characteristics of the wet natural carbonaceous granular material.

Belief that neither the soil nor groundwater chemistry was a factor in the corrosion process was supported by pH measurements. The pH of the insulation was as low as 4.8 whereas the pH of the adjacent soil varied from 7.7 to 9.0.

Conclusion

Based upon these results, it was concluded that the domestic hot water lines corroded because they had been exposed to an aggressive wet thermal insulation. The accelerated corrosion of the copper was caused by the presence of carbon and sulfur in a wet insulation, which had a low resistivity and a relatively low pH.

Hot Water Heating/Chilled Water Cooling System, Columbia, MO

A finely pulverized natural carbonaceous granular material was used to insulate the underground copper water tubes for a hot water heating/chilled water cooling system at a convalescent center constructed during 1974 in Columbia, MO. Approximately four years later, leaks developed in the copper tube that was exposed to a high water table. Examination of the copper tube revealed corrosion only where the tube had been in contact with wet insulation.

The corroded outside tube surface was covered with porous reddish-brown cuprous oxide which was, in part, overlain with a green copper corrosion product (Fig. 5). EDXA and microchemical analysis revealed that the green material contained appreciable quantities of copper along with minor amounts of sulfur and chloride, the overall corrosion products consisting primarily of copper sulfate and copper chloride.

Subsequent testing revealed that the thermal insulation did not contain any appreciable amounts of sulfur. It did however, have a relatively low pH value of 4.5. Unfortunately, sufficient insulation material was unavailable for resistivity measurements. The small amount of sulfur in this natural carbonaceous granular material was not considered unusual since subbituminous coal would be expected to contain varying amounts of this element.

FIG. 5—*External surfaces of corroded underground copper water tube that had been thermally insulated with a natural carbonaceous granular material. Typically, the corrosion products consisted of porous reddish-brown cuprous oxide overlain in part with green copper sulfate and copper chloride (~X1).*

Conclusions

Based upon the general guidelines established for copper water tube in underground building systems, it is apparent that finely pulverized natural carbonaceous granular insulation materials must be kept dry. If they become wet, they become poor thermal barriers and create an environment that is aggressive to copper and other commonly used construction materials. Copper cannot exhibit its normal resistance to corrosion if subjected to wet, low resistivity, low pH environments containing appreciable amounts of carbon, and sulfur.

Prevention Through Cathodic Protection

Fortunately, when metallic materials are inadvertently exposed to a hostile, underground, thermal insulation environment, they can be cathodically protected. Based in part upon the success of cathodic protection in mitigating the corrosion of copper tube heating lines exposed to wet insulating concrete [*8*], cathodic protection systems were designed and installed for the domestic hot water lines at the Brooklyn, NY, housing development.

The cathodic protection systems consisted of a deep anode bed and rectifier for each housing block at the development.[3] Deep anode beds were used be-

[3]Burke, N. D., private communication to J. R. Myers, 1 June 1983.

cause of the unavailability of open land for conventional anode systems and the desire to provide uniform current distribution to the underground piping. The cathodic protection systems concurrently provide protection to the uninsulated copper tube cold water lines and the coated steel gas service pipes in the area because all of the underground systems are electrically continuous.

Each of the anode beds contains eight vertically positioned, 76 mm (3 in.) diameter, 1524 mm (60 in.) long graphite anodes evenly spaced between 17 and 61 m (55 and 205 ft) below the surface. The anodes are backfilled in each 254 mm (10 in.) diameter, vertically drilled hole with low resistivity calcined petroleum coke breeze. Each anode bed is activated by an 80-V 34-A capacity direct current rectifier. Individual anode leads to a junction box for each rectifier; permanently installed reference electrodes and the necessary test stations were included in the cathodic protection design and installation to facilitate current control and system monitoring.

The results of field testing following initial energizing of the cathodic protection systems revealed that the thermally insulated copper water tubes could be sufficiently polarized for effective corrosion mitigation.

Corrosive Attack by Moist Cellulosic Insulation

Largely because of its low cost and general availability, cellulosic-type insulation has often been selected for reducing residential thermal losses. However, this type of insulation has a tendency to absorb moisture and, in the presence of certain chemicals added for fire retardation, may become corrosive to metals.

Cellulosic insulation is manufactured by shredding waste paper, usually newsprint, and then milling it into a finely divided fluffy fibrous material. To reduce its rather high flammability potential, fire retardant chemicals are added. The insulation is then bagged and shipped.

Quality control varies widely among the numerous companies that produce this material. In some cases, it has been reported[4] that the chemical additives have not been entirely absorbed, which has led to substantial residues from the paper making process at the mill being found in the bottom of the shipping container.

Fire retarding chemicals may or may not be corrosive. Borax, for example, may act as a mild corrosion inhibitor for metals like copper. On the other hand, aluminum sulfate is mildly corrosive and ammonium sulfate is extremely aggressive towards copper.

One detailed laboratory study by the authors [*9*], involving both failed copper plumbing tube specimens and the cellulosic insulation with which it was in contact, confirmed the presence of much ammonium sulfate with some indication of boron. As in all such cases where the insulation has been wet or a

[4]Communication from ASTM Committee C-16 on Thermal Insulation.

water extraction has been made, it is not possible to state with assurance whether the minor constituents were in the original paper stock or were added by the insulation manufacturer.

The copper and brass specimens in Fig. 6 showed patches of bluish material identified as partly ammonium sulfate and partly unknown compounds. Crystals of various intensities of blue color suggested the presence of Cu^{++} ions in an otherwise colorless crystal (such as ammonium sulfate). Small regions had the whitish cast sometimes seen when hydrates have lost some or all of their water.

Some hard blue material adhering to the pipe seemed to consist mainly of paper fibers, an unidentified binder, and a few tiny crystals. All of these adherent materials contained copper and boron in excess of 10% and minor amounts of such elements as silicon and aluminum.

Ash from the insulation remote from the pipe contained boron and aluminum in excess of 10% and minor amounts of calcium.

Table 2 [*10*] lists the compositions of fire retarding chemicals extracted from 19 different cellulosic insulation samples. The wide variation is noteworthy as is the great pH range of the water extract with one sample having a pH of only 3.7, which is distinctly acidic. As the actual formulations were not revealed by the chemical tests, there is no apparent correlation with pH and the chemical content.

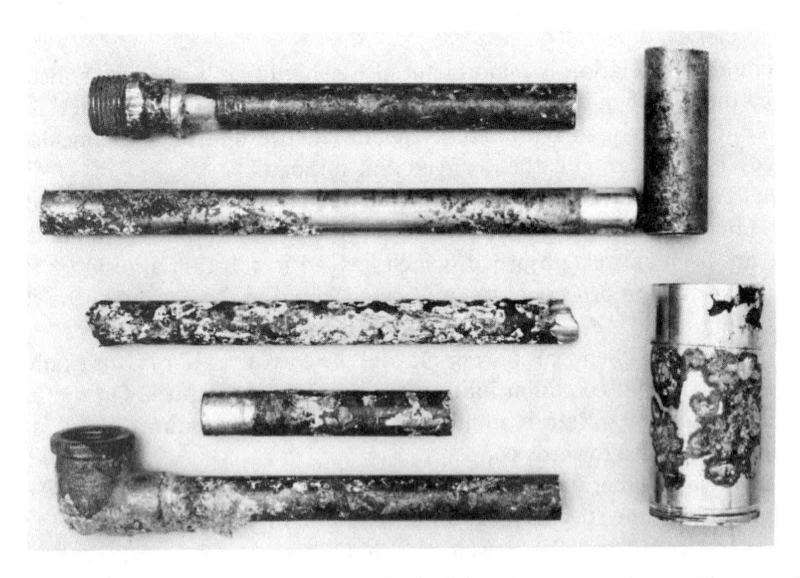

FIG. 6—*Outside surface of copper water tube displaying extensive attack caused by aggressive cellulosic insulation (~X1).*

TABLE 2—*Composition and pH of cellulosic insulation samples.*

			Fire Retardant Chemical, %				
Sample Identification	Total Water Solubles, %	pH	Ammonium Sulfate	Boric Acid	Calcium Sulfate	Aluminum Sulfate	Sodium Carbonate
526-1	18	4.4	18	...	...	...	...
526-5	20	8.0	...	11	1	...	5
527-A	22	8.2	...	16	...	...	3
527-B	31	4.8	...	23	...	...	...
527-C	28	8.1	...	22	...	...	5
527-C1	24	8.2	...	20	...	...	5
527-D	22	8.0	...	13	...	...	5
527-E	26	4.5	26	...	...	...	...
527-F	21	5.9	...	10	5	...	2
527-G	19	4.4	19	...	...	...	...
527-H	21	7.8	...	16	3	...	1
527-I	20	5.0	...	4	1	...	...
535	24	7.4	...	17	4	...	...
562	22	3.7	18	1	2	...	...
563-4	24	4.0	...	10	...	7	2
563-5	19	7.7	12	4	1	...	2
563-6	17	5.9	...	4	6	...	...
563-7	23	6.1	...	5	8	...	...
593	17	7.7	...	...	...	17	...

Revisions to Insulation Standards

Much unfavorable press notice [*11-15*] in addition to support from the U.S. Consumer Product Safety Commission [*16*] prompted the Federal Government to issue the much needed revision to Federal Standard Thermal Insulation (Loose Fill for Pneumatic or Poured Application): Cellulosic or Wood Fiber (HH-I-515), which was issued in 1979.

Concurrent ASTM Committee C-16 on Thermal Insulation action involves the revision of ASTM Specification for Cellulosic Fiber (Wood-Base) Loose-Fill Thermal Insulation (C 739) to contain a test method using flat copper specimens for corrosion evaluation under controlled conditions.

Conclusions

Cellulosic insulation under moist or humid conditions may not be a suitable material for use in contact with plumbing or electrical systems because the presence of contaminants and an added fire retarding chemical may make it potentially aggressive to copper.

This is complicated by the fact that moisture-free conditions cannot be assured in service, which, in turn, further activates the potential corrosive condition.

When moist insulation supports fungal growth, more moisture is taken into the system, and acid products resulting from biological processes may also cause corrosive attack.

References

[*1*] Myers, J. R. and Cohen, A., "Conditions Contributing to Underground Copper Corrosion," *Journal of the American Water Work Association*, Aug. 1984, pp. 68–71.

[*2*] Denison, I. A., "Electrolytic Behavior of Ferrous and Non-Ferrous Metals in Soil Corrosion Circuits," *Transactions of the Electrochemical Society*, Vol. 81, 1942, pp. 435–453.

[*3*] Gilbert, P. T., "Corrosion of Copper, Lead and Lead Alloy Specimens After Burial in a Number of Soils for Periods Up to 10 Years," *Journal of the Institute of Metals*, Vol. 73, 1947, pp. 139–174.

[*4*] Romanoff, M., "Underground Corrosion," National Bureau of Standards Circular 579, Washington, DC, 1957.

[*5*] McGannon, H. E., Ed., *The Making, Shaping and Treating of Steel*, Herbick and Iteld, Pittsburgh, PA, 1971, p. 78.

[*6*] Haslarn, R. T. and Russell, R. P., *Fuels and Their Combination*, McGraw-Hill Book Company, New York, 1926, p. 52.

[*7*] *Manual on Underground Corrosion in Rural Electric Systems*, Bulletin 161–23, Rural Electrification Administration (U.S. Department of Agriculture), Oct. 1977.

[*8*] Rogers, P. C., Gross, E. E., and Husock, B., "Cathodic Protection of Underground Heating Lines," *Materials Protection*, Vol. 1, No. 7, July 1962, pp. 38–43.

[*9*] *Corrosive Attack of Copper Water Tube by Cellulosic Insulation*, Copper Development Association Inc., Jan. 1978.

[*10*] *Survey of Cellulosic Insulation Materials*, Energy Research and Development Administration, U.S. Department of Commerce, Table 1, Jan. 1977, p. 4.

[*11*] "Push for Insulation Linked to Fire Peril," *The New York Times*, 27 Nov. 1977.

[*12*] "Insulation Hazards Mounting," *The Washington Post*, 1 Oct. 1977.

[*13*] *Energy Conservation Digest*, Vol. 1, No. 3, Washington, DC, 6 March 1978.

[*14*] *Energy Conservation Digest*, Vol. 1, No. 13, 24 July 1978.

[*15*] *Chemical Week*, 3 May 1978, p. 13.

[*16*] U.S. Consumer Product Safety Commission, Washington, DC, *Public Calendar*, Vol. 5, No. 43, 28 July 1978.

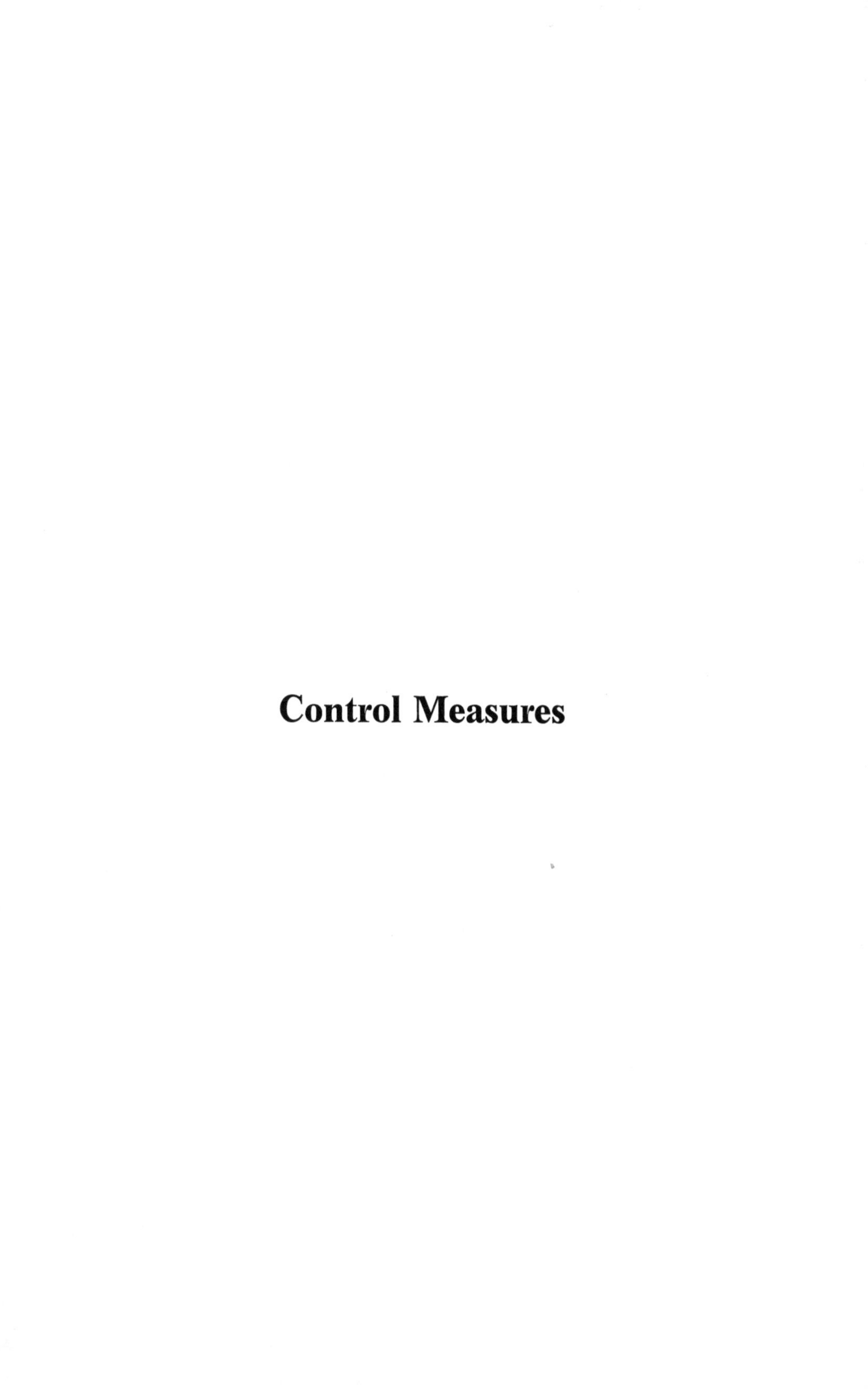

Control Measures

Paul E. Krystow[1]

Controlling Carbon Steel Corrosion Under Insulation

REFERENCE: Krystow, P. W., **"Controlling Carbon Steel Corrosion Under Insulation,"** *Corrosion of Metals Under Thermal Insulation, ASTM STP 880*, W. I. Pollock and J. M. Barnhart, Eds., American Society for Testing and Materials, Philadelphia, 1985, pp. 145-154.

ABSTRACT: Corrosion under hot and cold thermal insulation can be a serious problem in chemical plants. Not only has corrosion under insulation caused staggering maintenance costs in the millions of dollars but also has led to lost production time as well as affected plant safety. Many of the chemical plants have experienced a variety of problems and some of the specific experiences will be described in this symposium.

As a result, Exxon Chemical Company and many other chemical companies have embarked on a major effort to address the problem. In a companion paper, Exxon Chemicals at Baton Rouge will describe identifiable factors that cause the problem. This paper will review the important steps that Exxon Chemical has taken to control carbon steel corrosion under insulation. Specifically, the review will include (1) organizing and scheduling more rigorous programs of inspection including typical examples, (2) preparation of improved insulation specifications that address each of the factors affecting corrosion, and (3) action programs required to assure that improved insulation specifications are implemented.

An important aspect of controlling corrosion under insulation is through an appropriate inspection program. Unfortunately under normal circumstances, inspection requires the removal of the insulation during downtime, which is both costly and extremely difficult particularly on large towers and complex piping systems. There is an urgent need to develop a nondestructive onstream examination (NDE) method to detect corrosion without removal of insulation. In this connection, the Materials Technology Institute for the Chemical Process Industries is investigating some of the NDE methods. A brief synopsis will be made of the NDE methods employed. However, these methods have had only limited success, and a breakthrough inspection method is still needed. This need for breakthrough represents an important challenge to the international scientific community.

KEY WORDS: corrosion, insulation, carbon steels, thermal insulation

Corrosion under hot and cold thermal insulation can be a serious problem in chemical plants. Not only has corrosion under insulation caused staggering

[1]Engineering associate, Exxon Chemical Company, P.O. Box 271, Florham Park, NJ 07932.

maintenance costs in the millions of dollars, but it also has led to lost production time as well as affected plant safety. Many of the chemical plants have experienced a variety of problems and some of the specific experiences are being described in the other papers in this publication.

As a result, Exxon Chemical Company and many other chemical companies have embarked on suitable programs to address the problem. In a companion paper, Exxon Chemical Americas at Baton Rouge describes the identifiable factors that cause the problem.[2] This paper will review the important steps that Exxon Chemical has taken to control carbon steel corrosion under insulation from each of the identifiable factors.

Specifically, this presentation will review the following control measures as shown in Fig. 1:

- Prepare improved insulation specifications and guidelines for controlling corrosion under insulation.
- Organize and schedule more rigorous programs of inspection to assure that the improved insulation specifications are implemented.
- Visual inspection is the most reliable method, but some promising NDE methods are emerging.

Before discussing the items shown in Fig. 1, let us quickly review the various factors leading to corrosion under insulation as discussed in Lazar's paper (Exxon Chemical Americas, Baton Route).[2] In addition, we shall give examples of how these corrosion factors are overcome through improved insulation specifications and corrosion control guidelines. A brief synopsis of the various figures as reproduced in the attachment is presented below.

Figure 2

Equipment operating with a process temperature ranging between −3.89 to 121.11/148.89°C (25 to 250/300°F) is subject to corrosion under insulation, and consideration should be given to providing suitable paint protection underneath. If equipment operates at a temperature outside the range, then painting is not necessary. Equipment that operates alternatively from a temperature outside the range, but within the critical range for 20% of the time, should be painted. However, suitable high-temperature paints are required when elevated temperatures are encountered during alternate exposure within the critical temperature range. Recommendations on coatings are covered in several of the other papers in this publication.

Equipment geometry and design is very critical and Figs. 3 and 4 illustrate the variety of geometry configurations that can lead to corrosion problems.

Insulation type is important, not only from the cost and thermal insulation capability, but also in respect to moisture retention and water leachable con-

[2]Lazar, P., in this publication, pp. 11–26.

- PREPARE IMPROVED INSULATION SPECS AND GUIDELINES FOR CONTROLLING CORROSION UNDER INSULATION.
- ORGANIZE AND SCHEDULE MORE RIGOROUS PROGRAMS OF INSULATION INSPECTION.
- VISUAL INSPECTION IS MOST RELIABLE BUT SOME PROMISING NDT METHODS ARE EMERGING.

FIG. 1—*Controlling carbon steel corrosion under insulation.*

- PROCESS TEMPERATURE
  - 25°F TO 300°F
  - NORMAL AND ALTERNATE OPERATIONS
- EQUIPMENT GEOMETRY AND DESIGN
  - WET EXPOSURE DURATION
  - FREE DRAINAGE ESSENTIAL
  - AVOID WET SPOTS
- INSULATION TYPE
  - THERMAL PROPERTIES
  - COST
  - WATER LEACHABLE CONTENTS
  - WATER RETENTION, DRYING

FIG. 2—*Factors affecting corrosion under insulation.*

tents. Insulation types, such as calcium silicate, which can absorb and retain substantial amounts of water and leach chlorides, can lead to more problems than other insulations such as foam glass, that is not normally subject to moisture absorption and does not contain significant leachable material. For critical insulation service, it is necessary to consider all the factors, including those relating to corrosion.

Figures 3 and 4

Figures 3 and 4 illustrate the equipment geometry and the variety of attachments that appear on a vessel, the points where water can enter and become trapped and the areas where corrosion can occur underneath insulation. It is recommended that similar guideline sketches be prepared for all types of equipment, that is, piping, storage tanks, exchanger shells, drums, and so forth, to facilitate proper inspection and serve for preparing appropriate mechanical design sketches for the attachments that will provide more adequate weatherproofing, vapor proofing and painting.

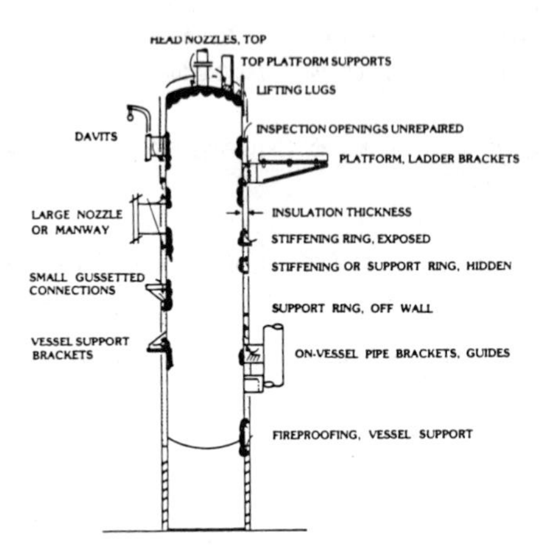

FIG. 3—*Inspection of vessels ambient to 121°C (250°F) for CUI.*

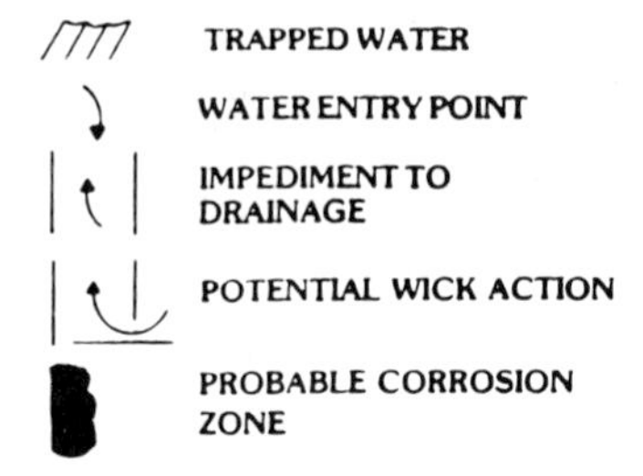

FIG. 4—*Inspection of vessels ambient to 121°C (250°F) for CUI.*

Some undesirable geometry and design features that should not be used are as follows:

(1) flat horizontal surfaces (vacuum rings),
(2) structural shapes that trap water ("H" beams),
(3) shapes that are impossible to properly weatherproof (structural shapes, gussets, and so forth),
(4) shapes that funnel or lead water into the insulation (angle iron brackets), and
(5) inadequate spacing that causes interruption of the vapor or weather barrier (nozzle extension, ladder brackets, and deck spacing).

A good discussion of equipment design is included in Lazar's paper.[2]

Figure 5

This sketch details in Case A, a possible method for overcoming corrosion from exposed carbon steel ledges as may exist with vacuum rings that can

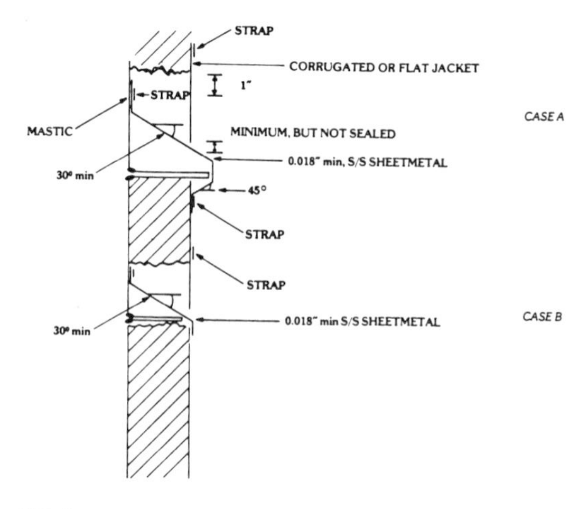

FIG. 5—*Insulation at stiffening/old insulation support rings.*

allow water to wick upward into the insulation. Case B illustrates a method that allows for drainage of water that may enter through breaks in the weatherproofing/vapor barrier, rather than the water collecting at the various impediments imposed by the attachments, support rings, and so forth.

Figure 6

As a secondary defense to preventing corrosion at service temperatures from −3.89 to 121.11/149.89°C (25 to 250/300°F), it is recommended that painting be applied to protect the steel under insulation. Not only should the specification include type of paint, surface preparation, inspection requirements, and so forth but also should be satisfactory for all temperatures that may be encountered, particularly during upset operation.

The design of weatherproofing/vapor proofing is considered to be the primary defense against corrosion and needs special attention around attachments and protrusions, such as illustrated in Fig. 3. The use of proper sealants around protrusions is especially important and needs to be checked and reapplied in critical areas at least once every two years. Further specific details are included in Lazar's paper.[2]

Climatic conditions are extremely important, and locations, such as the Gulf Coast, where humidity and rainfall are high, require more attention than in the less humid northern areas such as in Canada. Also, insulated equipment in the vicinity of cooling towers or subject to splash from further washing operations require special attention. Equipment operating at temperatures below the dew point, which is subject to frequent sweating, is also of special concern. Thus, the climatic condition, location, and operating condi-

- PAINT SELECTION
  - SECONDARY DEFENSE
  - EXTENDS SYSTEM LIFE
- WEATHER/VAPOR PROOFING
  - PRIMARY DEFENSE
  - PROTECTS AGAINST PHYSICAL, LIQUID AND VAPOR
  - CLOSURES AROUND PROTRUSIONS
  - MASTIC LIFE LIMITING
- CLIMATE
- MAINTENANCE PRACTICES
  - REPAIR INSULATION DAMAGE
  - INSPECTION PROGRAM

FIG. 6—*Contributing factors.*

tion will influence whether more or less stringent insulation specifications and inspection programs are required.

Maintenance practices are extremely important. All equipment is eventually exposed to insulation damage, and a rigorous repair program is necessary. The insulation repair program should not only detail the proper maintenance procedure but include inspectors participation to verify that the insulation has been properly installed. This is particularly important since insulation work is often carried out by contract personnel who may not be familiar with your insulation repair specifications. Often it is necessary to complete maintenance insulation while onstream. This can be particularly difficult with cold service insulation systems that are subject to icing. Procedures have been developed using methanol spray wrapping with polyethylene sheet and taping to prevent ice from reforming before insulation installation.

Figure 7

This figure illustrates a natural tendency that occurs with insulation maintenance. Not only is finishing the job important from an energy conservation standpoint, but it would be even more important from the corrosion standpoint if the same system as illustrated were upside down and rainwater and atmospheric moisture could enter and collect into the piping system.

Figure 8

It is important to initiate an organized plant-wide program for controlling corrosion under insulation. This is best accomplished through preparation of

FIG. 7—*The natural tendency that occurs with insulation maintenance.*

- DESIGN FEATURES
- COMPLEX GEOMETRY DETAILS
- INSULATION MATERIALS
- PAINT SELECTION
- WEATHER/VAPOR PROOFING
- PROJECT ENGINEER GUIDELINES
- INSPECTION PROCEDURES AND CHECKLISTS
- MAINTENANCE REPAIR TECHNIQUES (INCLUDING ON-STREAM REPAIR PROCEDURES)

FIG. 8—*List of what the guidelines cover.*

a guideline that addresses all the concerns and pitfalls. Figure 8 lists the important items that should be included in the guidelines, and we suggest the preparation of similar guidelines that conform with the particular needs of your plant.

Figure 9

An important aspect of controlling corrosion under insulation is through an appropriate inspection program. Unfortunately, under normal circum-

- VISUAL
- ULTRASONICS AND MICROWAVE
- RADIOGRAPHY
- TANGENTIAL RADIOGRAPHY
- X-RAY WITH IMAGE INTENSIFICATION AND CLOSED-CIRCUIT TELEVISION
- INFRARED
- CONDUCTIVITY MEASUREMENTS
- POTENTIAL MEASUREMENTS
- ELECTROMAGNETIC METHODS USING HALL SENSORS

FIG. 9—*Current inspection methods employed.*

stances, inspection requires the removal of the insulation during downtime, which is both costly and extremely difficult, particularly on large towers and complex piping systems. There is an urgent need to develop a reliable nondestructive onstream examination (NDE) method to detect corrosion without removal of insulation. In this connection, the Materials Technology Institute for the Chemical Process Industries is investigating some of the NDE methods. There have been some promising methods under investigation which are discussed in more detail in other papers in this publication. Figure 9 lists some of the current inspection methods employed.

Figure 10

At the moment, the most reliable, positive inspection method is by direct visual examination that requires physical removal of the insulation. It is difficult to pinpoint areas to be checked, but, most assuredly, areas where wet

- INSPECTION CONDITIONS OF WEATHER PROOFING EVERY TWO YEARS PARTICULARLY AFTER TURNAROUNDS
- CHECK METAL JACKETNG EVERY TWO YEARS PARTICULARLY AFTER TURNAROUNDS
- REMOVE INSULATION IN SELECT AREAS OF THE CRITICAL WET ZONES EVERY FIVE YEARS AND INSPECT
- REPAIR AND REINSULATE AFTER INSPECTION SHALL BE WITNESSED BY INSPECTOR

FIG. 10—*Suggested guidelines for inspection of corrosion under insulation.*

- ESTABLISH MORE RIGOROUS PROGRAM OF INSPECTION
- PREPARE IMPROVED INSULATION SPECIFICATIONS WHICH MINIMIZE THE FACTORS CAUSING CORROSION
- INITIATE PROGRAM TO ASSURE ITEMS ABOVE ARE CARRIED OUT EVEN IF COSTLY
- PROMOTE NONDESTRUCTIVE METHODS FOR DETECTING CORROSION UNDER INSULATION - BREAKTHROUGH NEEDED

FIG. 11—*Program for overcoming corrosion under insulation.*

insulation exists is where the corrosion will take place. These areas tend to be at lower locations of equipment and piping where moisture can accumulate and collect at obstructions. Also, equipment is subject to sweating caused by service conditions or located in a vicinity where cooling water drift or water washing splash can be encountered.

The cutting of small insulation windows for spot checking corrosion under insulation is not always adequate. It is necessary to remove large areas of insulation in critical wet regions and realize that there is a level of risk where one has not looked. However, it is emphasized that dry areas of insulation do not experience corrosion, and visual inspection need not be carried out in these areas. Figure 10 gives a suggested guideline for inspection.

Figure 11

This is the concluding slide. It summarizes the program for overcoming corrosion under insulation as presented in this paper.

DISCUSSION

A. S. Krisher[1] *(written discussion)*—This conference has been a very useful and stimulating review of a widespread and serious problem. It is apparent that a control program consisting of coating before insulation, proper design of the system, proper installation of the system, thorough periodic inspection,

[1]Monsanto Co., 800 N. Lindbergh-CS6F, St. Louis, MO 63167.

and proper maintenance will substantially reduce the occurrence of corrosion under insulation.

It is worth noting that most of the steps listed above are identical to those required to assure optimum energy conservation. Consideration of this subject may help to justify improved practices in cases where corrosion control alone is judged to be insufficient justification for the expense of such practices.

A critical subject that was largely overlooked in this conference is the question of economics. In my opinion, we do not have a reasonable basis for estimating the probable costs if we do nothing to prevent the problem. There are many cases where corrosion under insulation has necessitated very extensive and expensive repairs. However, in a number of these cases, repairs were not required until long after the design life of the plant had been exceeded. There are also a large number of cases in which no preventative steps were taken and no problems were encountered.

I believe that we need to derive a reasonable basis for estimating the probability, cost and timing of corrosion under insulation. This could then be used to evaluate the justification of various corrective steps. I have a subjective feeling that we may sometimes spend more time on preventative measures than is justified.

Paul E. Kryston (author's closure)—The comments offered by Bert Krisher are appropriate, but the response is difficult to explain adequately. It is agreed that the economics for CUI has been somewhat overlooked in this conference, but not necessarily neglected by the chemical companies concerned with CUI. Unfortunately, the costs for unexpected downtimes and the money spent for conducting appropriate inspection and maintenance needed to make an economic justification are usually considered to be proprietary and therefore cannot be discussed.

We are in total agreement with the statement that it is necessary to establish a reasonable basis for estimating the probability, cost, and timing of corrosion under insulation. Most of the equipment in a chemical plant is not subject to corrosion under insulation. This is why special emphasis was made (see Fig. 2 and in Exxon Chemical's companion paper by Lazar) of the factors leading to CUI.

It is important to review the factors affecting CUI to establish the probability and the timing of CUI, and together with an evaluation of the costs of unexpected downtimes and the costs to carry out maintenance/inspection efforts, determine the economic justification of the corrective program. In many petrochemical processes unexpected shutdowns can lead to substantial production losses, and this is a key concern along with safety in establishing the corrective action program. On the other hand, there are also many chemical processes where a leak caused by CUI does not lead to a serious safety concern and a shutdown does not involve significant production losses so that corrective action beforehand is not a vital criteria.

Peter A. Collins,[1] *John F. Delahunt,*[1] *and Debbie C. Maatsch*[1]

Protective Coating System Design for Insulated or Fireproofed Structures

REFERENCE: Collins, P. A., Delahunt, J. F., and Maatsch, D. C., **"Protective Coating System Design for Insulated or Fireproofed Structures,"** *Corrosion of Metals Under Thermal Insulation, STP 880,* W. I. Pollock and J. M. Barnhart, Eds., American Society for Testing and Materials, Philadelphia, 1985, pp. 155–164.

ABSTRACT: Corrosion of carbon steel equipment beneath thermal insulation and concrete is a critical concern among refineries, chemical plants, marketing organizations, and pipelines. Corrosion has been detected worldwide under all types of thermal insulating materials. At Exxon this concern has been addressed in a number of laboratory investigations and also during field troubleshooting assignments. This paper describes work undertaken to develop protective coating systems suitable for equipment to be insulated or fireproofed. These systems are described. In addition, the advantages of organic coatings compared to inorganic zinc-rich coatings are reviewed. Also discussed are recently completed novel evaluations to select coating systems for thermally insulated and thermally cycled equipment.

KEY WORDS: corrosion, insulation, tests, paints, coatings, steels

Corrosion of carbon steel equipment beneath thermal insulation and concrete is a critical concern among refineries, chemical plants, marketing organizations, and pipelines. Corrosion has been detected worldwide under all types of thermal insulating materials. At Exxon this concern has been addressed in various laboratory investigations. This paper describes work undertaken to evaluate protective coating systems suitable for equipment to be insulated or fireproofed. Discussed are recently completed novel laboratory evaluation techniques to select coating systems for insulated and thermally cycled equipment. In addition, the advantages of organic coatings compared to inorganic zinc rich coatings (IOZR) are reviewed.

[1]Staff engineer, engineering associate, and project engineer, respectively, Exxon Research and Engineering Co., P.O. Box 101, Florham Park, NJ 07932.

Corrosion Under Insulation Widespread

Corrosion under insulation and fireproofing has occurred in all major areas of operation and beneath all major types of insulation in use. These problems began to come to the fore during the late 1960s and became of major concern in subsequent years. A major contributing factor for this corrosion problem was changes made to materials specified for insulation systems previously:

- *Insulations*—New cellular plastic foams (polyurethane and phenolic) came into use. When water saturated, these are acidic and extremely corrosive.
- *Metal Jackets*—Reduced anywhere from 50 to 100% in thickness. Composition changed from copper-bearing carbon steel to carbon steel. Zinc coating thickness reduced. Such changes would lead to wind damage and early corrosion failure.
- *Coating Jackets*—Many systems changed, for example, hot tar three-ply roof systems to, in many cases, 0.5 to 1.0 mm (20 to 40 mil) of synthetic rubber for storage tank roofs. Long-term weather resistance would be reduced.
- *Equipment Painting*—At specific locations, equipment painting reduced the need to paint equipment operating between 0 to 93°C (30 to 200°F) to between 0 to 55°C (30 to 130°F) before insulation application. This change increased the opportunity for corrosion to occur.
- *Fireproofing*—Eliminated external painting of concrete and also steel-concrete seam protection. Therefore, water could penetrate the concrete more readily and increased the corrosion potential.

These factors, along with external influences, such as increased use of recirculating salt water cooling towers, have led to increased corrosion under insulation and fireproofing.

Exxon over recent years has conducted programs to mitigated such corrosion. These included several evaluations of paints and protective coatings that would be applied to carbon steel equipment or structures before insulating or fireproofing. This paper describes the following investigations:

- Evaluation of paints and coatings under wet insulation or fireproofing.
- Determination of the bond strength of concrete applied to coated and uncoated steel.
- High-temperature cyclic tests of paints and coatings under wet insulation.

These programs are discussed in subsequent paragraphs.

Coating System Design for Insulated Equipment

Paints and coatings applied to insulated equipment may be designed to withstand immersion in hot water (93°C [200°F]) on a continual basis. However, such a design would be inordinately expensive since minimum surface preparation for such a service would be Steel Structures Painting Council-5 (SSPC-5), white abrasive blast cleaning, followed by the application of an expensive, high build sophisticated coating system. A more practical approach, based upon our experience, is the use of SSPC-6, commercial abrasive blast cleaning, and the application of a more easily applied, general maintenance paint or coating. Based upon such a philosophy, several laboratory screening evaluations were undertaken at steady state and cycling high-temperature conditions.

Evaluations of Coatings Under Insulation/Concrete

The evaluation of paints and coatings for use under thermal insulation and concrete consisted of two parts:

- Comparison of bare and coated steel corrosion coupons after 30-days exposure to wet insulation and concrete at 66°C (150°F).
- Determination of the effect of coatings on the bond strength of steel to concrete.

For the exposure tests, carbon steel coupons 100 by 100 by 3 mm (4 by 4 by 1/8 in.) were prepared by sandblasting and coating. The coatings evaluated were an inorganic zinc-rich coating (IZRC), a red lead alkyd, lead chromate phenolic, and an amine-cure epoxy. Each of the coated samples was scribed to introduce a defect in the coating. To simulate service exposure, the prepared samples were each encased in a "sandwich" of insulation (fire retardant polyurethane foam and calcium silicate) or concrete (Type 1 portland cement) as shown schematically in Fig. 1. The coated test panels were exposed

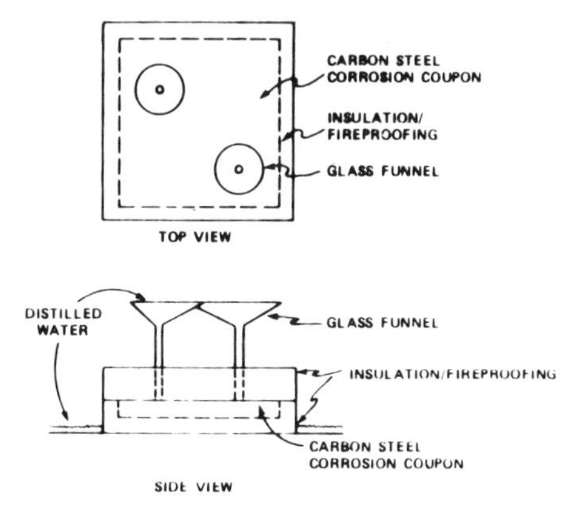

FIG. 1—*Specimen used to measure the corrosivity of carbon steel beneath wetted thermal insulation and concrete fireproofing.*

to an alternating wet and dry cycle using distilled water for 30 days. Temperatures were maintained at 66°C (150°F). Uncoated carbon steel test coupons without the covering of insulation or fireproofing were also tested for comparison.

All Coatings Except IZRC Perform Well

After the 30-day exposure, the coupons were removed and visually examined for rusting, discoloration, chalking, pitting, and other evidence of coating degradation. The steel coupons exposed without coatings exhibited varying degrees of corrosion—from light rusting under cement to heavy rust buildup when tested under polyurethane foam. The corresponding corrosion rates calculated from weight loss measurements ranged from 0.025 to 0.050 mm/year (1 to 2 mpy) for the sample tested under concrete to 0.175 to 0.200 mm/year (7 to 8 mpy) under the fire retarded polyurethane foam. Corrosion pitting was also evident on these samples after the corrosion scale was removed. The corrosion rate of the control sample was 0.050 to 0.075 mm/year (2 to 3 mpy).

The coated samples were generally in good condition following the test. There was no evidence of film breakdown except for the IZRC as discussed below. There was essentially no rusting, except for the samples tested under polyurethane foam, which exhibited rapid corrosion of the steel exposed by the scribe marks. Weight measurements of the painted samples before and after the test indicated negligible weight change.

Because of a slight pitting on the surface of the IZRC coated coupons, cross sections of these samples were examined metallographically. It was evident that the coatings were undergoing attack and that corrosion products were building up at the steel/coating interface. A comparison of an unexposed IZRC sample to the test coupons is shown in Fig. 2.

The unexposed coating is tightly adherent to the steel substrate and contains only a few small pores typical of this material. The exposed coatings showed areas of disbonding, severe porosity, and underlying corrosion. The corrosion products were qualitatively analyzed and found to contain silicon and iron oxides. While attack of the IZRC was evident after testing under all insulations and fireproofing, this degradation was most pronounced under the polyurethane foam insulation.

Calculated corrosion rates for the inorganic zinc coating based on weight loss, ranged from 0.050 to 0.090 mm/year (2 to 3.5 mpy) beneath polyurethane foam and calcium silicate, to 0.025 mm/year (~1 mpy) under cement. However, because these were calculated based upon the density of zinc, they are conservative because of the voids present in an IZRC film. Complete penetration of the IZRC film would occur within approximately 1 to 3 years (based upon the density of zinc).

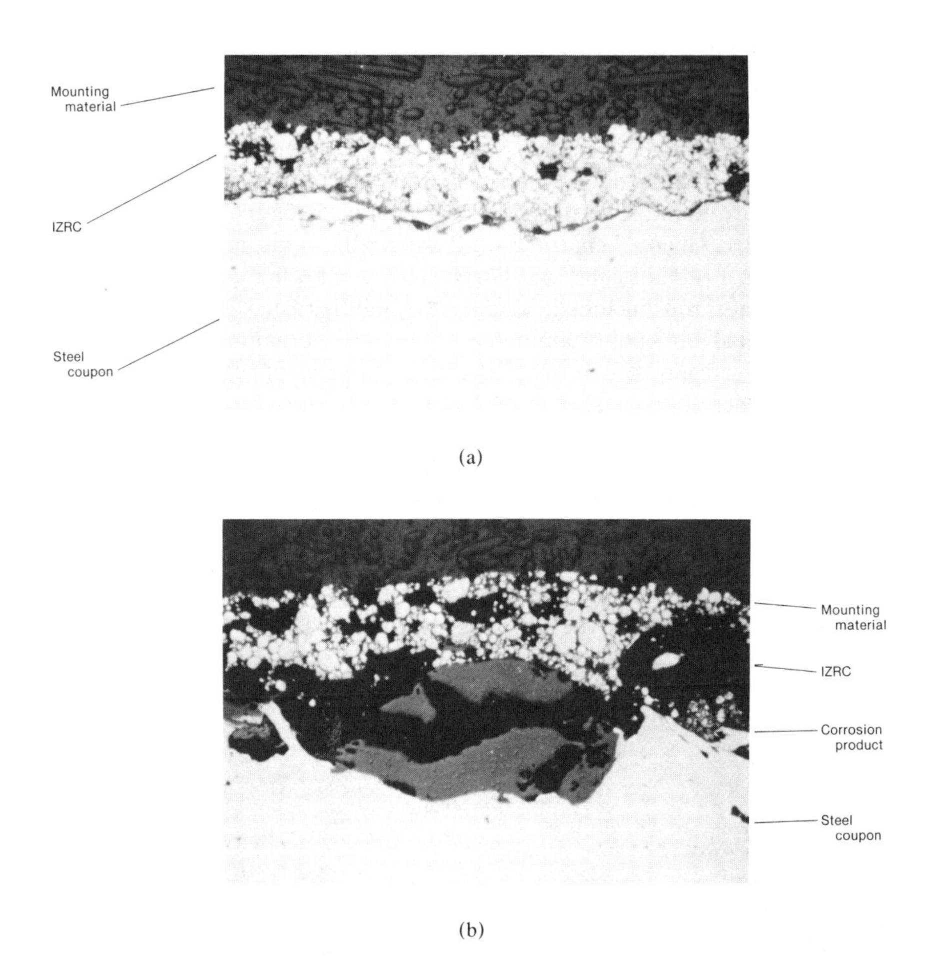

FIG. 2—(a) *Cross section of a panel coated with IZRC before testing. The coating is continuous, tightly adherent to the steel substrate, and contained a few isolated pores.* (b) *Cross section of a panel coated with IZRC after testing under polyurethane foam. Degradation of the coating has occurred with a buildup of corrosion products at the coating/substrate interface. (X150)*

Effect of Coatings on Bond Strength Examined

In the case of concrete applied to structural steel, an additional consideration may be the effect of an applied coating on the adhesion or bond of the steel to concrete. Bonding is important for the mechanical integrity of reinforced concrete structures. Also, well bonded concrete will prevent the ingress of water and other contaminants that will lead to the corrosion of the underlying steel. Various standards now specify that concrete must be applied directly to sandblasted steel. The reason for this is that concrete will not adhere as well to painted steel as to uncoated steel. Therefore, limited tests were undertaken to quantify the degree of bonding of steel with and without coatings.

To investigate bond strengths, specimens were prepared in accordance with ASTM Test Method for Comparing Concretes on the Basis of the Bond Developed with Reinforcing Steel (C 234). The specimens consisted of a carbon steel rod 1.9 by 45.7 cm (0.75 in. diameter by 18 in. long) encased in a 15.2-cm (6-in.) cube of concrete. The surface preparation and coatings applied to the steel rods included: sandblasted, lightly rusted, rusted in the atmosphere, coal tar epoxy, amine-cured epoxy, inorganic zinc, red lead primer, and lead chromate primer. The coatings were applied to sandblasted rods. The concrete was mixed to ASTM Making and Curing Concrete Test Specimens in the Laboratory (C 192) using a cement/aggregate/water ratio of 2:3:1 by weight. The samples were cured for 30 days. The specimens were tested by measuring the load required to break the bond of the steel rod to concrete and pull it from the cement cube using a standard tensile machine as shown in Fig. 3. The results of these pullout tests are shown in Fig. 4.

The best bond, as expected, resulted from the sandblasted bar, requiring 19.5 N (4400 lb) to break the concrete/steel bond. This bond strength was

FIG. 3—*Apparatus used to measure the bond strength of coated and uncoated steel bars to concrete. The concrete cube was held fixed by the upper arm of the tensile machine, while the lower arm pulled the bar downward. Loads were monitored to the nearest 89 N (20 lb).*

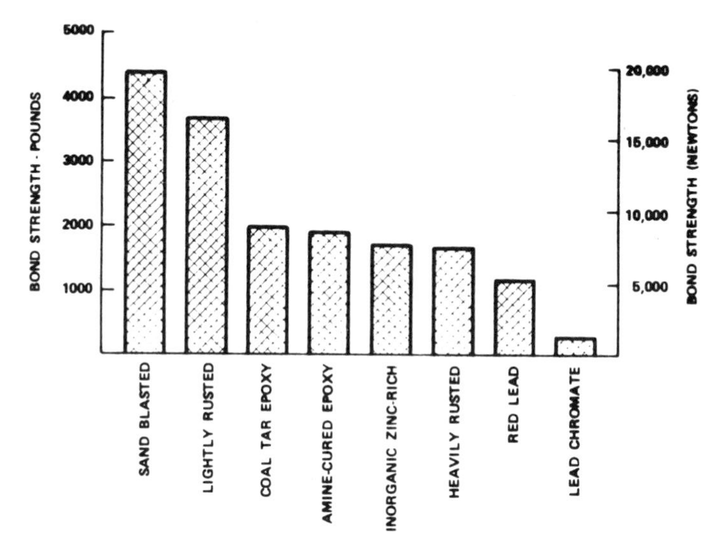

FIG. 4—*Pull out test results.*

slightly reduced by a rust blooming of the sandblasted steel and markedly reduced by atmospheric rusting or when concrete is applied to coated steel. The coatings that produced the highest bond strengths were the two epoxies and the inorganic zinc. It is somewhat surprising that the highest bond strengths resulted from the epoxy coatings that have glossy finishes, whereas the flat primers gave lower bond strengths. However, the results with epoxy based coatings are supported by similar tests by other investigators [*1*].

Although the bond strengths of coated steel rods to concrete were less than half that of sandblasted steel rods to concrete, the tests show that a bond does exist between coated steel and concrete. While this work did not identify a minimum bond strength necessary for acceptable field performance, corrosion of the underlying steel should not occur as long as the coating remains intact and there is no loss of adhesion between the steel, coating, and cement.

High-Temperature Cyclic Evaluation Undertaken

High-temperature cyclic testing of paints and coatings was undertaken because of the severity of corrosion occurring on insulated equipment in this type of intermittent service in operating plants. Testing conditions sought to duplicate "worst case" operating conditions. Paints were exposed to temperatures ranging from ambient to 315°C. This severe temperature range was used to accelerate coating failure.

The high-temperature test equipment is illustrated in Fig. 5. The unit was a modification of the apparatus used previously to evaluate heat resistant paints and also, temperature indicating paints. Insulated carbon steel pipes were suspended several inches over the heat source (propane burner). Chrome-alumel thermocouples were directly attached to the insulated pipe, and in conjunction with a recorder, supplied a continuous temperature profile of the pipe. A representative temperature profile is shown in Fig. 6.

The test procedure was designed to simulate cyclic service beneath wet insulation. During each test period, the pipe was heated for 8 h, then left at ambient temperature for the remainder of the day. Water was added daily to the insulation-pipe interface. Approximately 1 L (1 qt) was applied at the top of the pipe and allowed to run down the full length of the pipe. A basin welded onto the pipe bottom prevented any excess water from running into the heat source and created a reservoir to keep the insulation moist. The water entirely evaporated by the end of each 8-h heating period. Each pipe was cycled over a 30-day period.

Insulation selection was also based on increasing the severity of the test. Calcium silicate was used to insulate the lower two-thirds of the pipe, which was the portion exposed to the higher temperatures. Calcium silicate was cho-

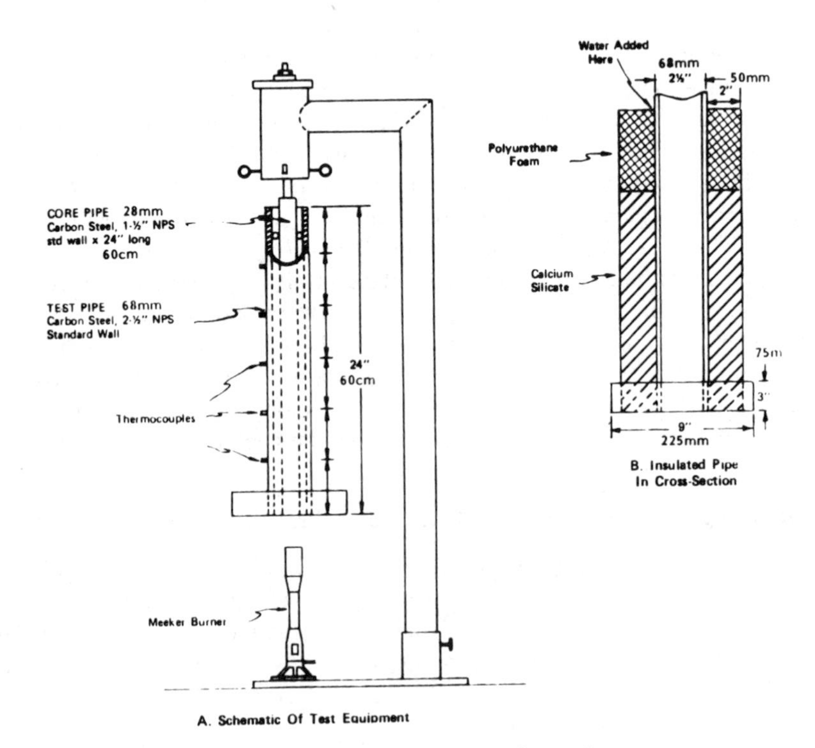

FIG. 5—*High-temperature paint testing equipment.*

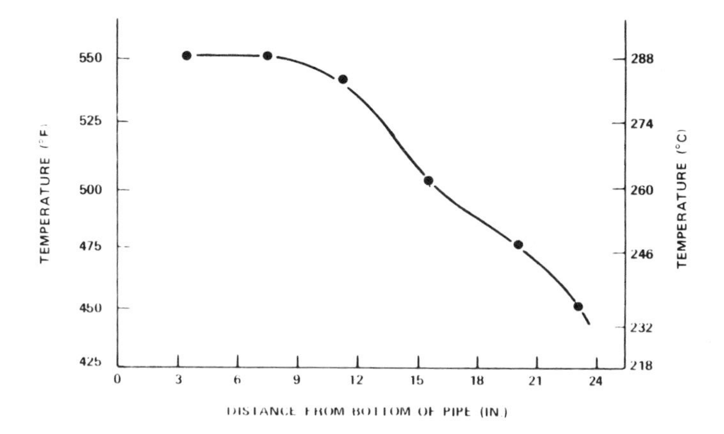

FIG. 6—*Temperature profile of test pipes.*

sen for two reasons: it is frequently used by refineries, and it is easy to handle. In addition, water leached through calcium silicate is known to be alkaline and thereby increases the severity of the test. Fire retardant polyurethane foam was used to insulate the upper one-third of the test pipe, even though the pipe surface temperature exceeded the suggested operating limit for polyurethane foam (83°C [180°F]). However, polyurethane foam was included because earlier work showed that water leached from this type of insulation, when treated with fire retardants, was more corrosive because of an acidic pH than other frequently used types of insulation. Therefore, it was felt that this would provide a severe test of the protective properties of the paints and coatings.

Six Coatings Evaluated

Six coatings were evaluated by the described test method. These included: phenolic lead chromate primer, catalyzed epoxy amine primer, zinc-pigmented linseed oil-based paint, zinc-pigmented silicone acrylic primer, and two epoxy phenolic coatings. These coatings were selected for several reasons. The phenolic and amine epoxy are the most commonly used organic coatings for protection beneath insulation. Although there is little knowledge concerning their performance in high-temperature cyclic service and the test temperature exceeded their suggested operating limit, they were included to provide a basis for evaluating other coatings caused by extensive experience with their performance. The silicone acrylic is heat-resistant paint, and it is currently specified for use beneath insulation by some refineries. The oleoresinous coating was chosen for its temperature resistance and ability to be applied over a lesser degree of surface preparation, unlike most of coatings tested. The two

TABLE 1—*High temperature cyclic testing.*

Paint System	Calcium Silicate	Polyurethane Foam
Epoxy phenolic	destroyed	accelerated corrosion
Epoxy phenolic	chalky, nonadherent	severe attack
Phenolic chromate	cracked, flaked, 30% destroyed	destroyed
Amine epoxy	extensive flaking	destroyed
Oleoresinous	intact	thin rust scale
Acrylic silicone	intact	limited protection

phenolic-epoxy coatings were selected for their outstanding chemical resistance and water resistance.

Results Confirm Severity of Environment

Results of the high-temperature cyclic testing confirmed the severity of the environmental conditions represented by this test. A summary of coating performance is shown in Table 1. Of the coatings tested, only the oleoresinous and acrylic silicone can be specified for use to protect equipment that may be thermally cycled. The other types of paints and coatings failed to some extent beneath calcium silicate insulation and almost catastrophically failed beneath fire retarded polyurethane foam.

Conclusions

From these evaluations, we have reached the following conclusions:

- Many organic maintenance primers should give satisfactory service beneath insulation and concrete. Lead chromate phenolics and amine epoxies have been used satisfactorily in the field.
- IZRC primer should not be used under insulation or concrete. Field experience also supports that conclusion.
- For high-temperature cyclic service under insulation oleoresinous or silicone acrylic coatings are recommended based upon accelerated testing.

Reference

[*1*] Mathey, R. G. and Cligton, J. R., "Bond of Coated Reinforcing Bars in Concrete", *Journal of the Structural Division*, Proceedings of the American Society of Civil Engineers, Jan. 1976.

Louis C. Sumbry[1] *and E. Jean Vegdahl*[2]

Prevention of Chloride Stress Corrosion Cracking Under Insulation

REFERENCE: Sumbry, L. C. and Vegdahl, E. J., **"Prevention of Chloride Stress Corrosion Cracking Under Insulation,"** *Corrosion of Metals Under Thermal Insulation, ASTM STP 880*, W. I. Pollock and J. M. Barnhart, Eds., American Society for Testing and Materials, Philadelphia, 1985, pp. 165–177.

ABSTRACT: Chloride stress corrosion cracking of stainless steel under insulation can occur because of concentration of chlorides at the metal surface. This paper discusses our company's current recommended practices for controlling this type of corrosion and results of a research program to improve our recommended practices.

Our current recommended practice for controlling corrosion of stainless steel under insulation is to select insulation materials low in extractable chlorides, jacket or coat insulation to exclude water and chlorides from the environment, and to coat the stainless steel equipment where applicable. Coating stainless steel to prevent contact with chlorides is recommended with some reservation because of generally poor experience in coating stainless steel equipment.

A research program is being conducted on coatings for stainless steel equipment. A variety of coatings have been evaluated using adhesion and flexibility tests, outdoor exposure, salt fog cabinet, environmental cabinet containing hydrogen sulfide, carbon dioxide, nitrogen dioxide, and wet air, a Weatherometer cabinet, elevated temperature exposure (260 to 500°C), thermal shock tests, and chloride permeability tests. Adhesion to stainless steel is generally fair. For elevated temperature applications, straight silicones are generally superior. Many coatings are adequate for ambient temperature applications if protected from mechanical damage.

KEY WORDS: chlorides, stress corrosion, insulation, jackets, coatings, stainless steels, mechanical bonds, chemical bonds

Chloride stress corrosion cracking of insulated stainless steel equipment is a particular problem because chlorides tend to concentrate under the insula-

[1] Senior chemist, Gulf Research and Development Co., P.O. Drawer 2038, Pittsburgh, PA 15230.

[2] Consultant, 14120 SW Rochester Dr., Beaverton, OR 97005; formerly, engineer, Gulf Research and Development Co., Pittsburgh, PA 15230.

tion at the surface of the metal if the insulation becomes wet. The moisture can leach soluble chlorides out of the insulation, or may already contain chlorides from the environment. At the warm metal surface the moisture is vaporized, leaving behind an increasing concentration of chlorides. Then during operation or startup/shutdown when the piping is in the susceptible temperature range, chloride stress corrosion cracking can quickly take place. Our experience indicates that the problem is primarily within the 50 to 200°C (120 to 390°F) temperature range. In this temperature range, waterborne chlorides concentrate by evaporation of the water.

The exact mechanism of chloride stress corrosion cracking is still a matter of discussion, however, the conditions required for it to take place are clear and fairly well known. The four required factors are an austenitic stainless steel, chloride ions, tensile stress on the metal, and certain temperature limits. The cracks are generally highly branched, transgranular, and occur with no deformation, and hence no warning.

This paper will discuss our current recommended practice for controlling corrosion of stainless steel under insulation and results of a research program to improve our recommended procedures for coating under insulation. As this paper is concerned with chloride stress corrosion cracking, we will only deal with equipment operating at 50°C (120°F) and above.

Current Insulating Practice

The following is a discussion of the insulation practices that have been in use at Gulf over the last 10 to 15 years. These include the design standards, insulation and accessory materials, and installation practices. The final section of this paper discusses work that is being performed to update these practices.

Design Basis

Unless a heat loss is desired, vessels, heat exchanger units, and piping operating at temperatures above 93°C (200°F) are insulated. Insulation at lower temperatures is required when it is necessary to limit heat losses or for personnel protection.

Codes—The basic design, assembly, and installation must conform to federal, state, and local codes or regulations. If a conflict exists among the codes and regulations, the most stringent requirement will govern.

Insulation Thickness—The economic thickness, based on Gulf's standard drawings for rigid type insulation, is defined as the thickness of a given insulation that will save the greatest amount of energy while paying for itself within an assigned period of time.

Materials

Insulation—The high-temperature insulation materials listed in Table 1 are used for vessels, heat exchangers, pumps, and piping, with the exceptions that are listed below:

- No asbestos bearing insulation materials (except mastic or asphalt base materials) are used in new installations or as replacement material in existing plants.
- The insulating materials used on austenitic stainless steel surfaces are low in extractable chlorides (<50 ppm). For additional protection against chloride attack, austenitic stainless steel surfaces operating below 204°C (400°F) are painted with a zinc free silicone or an epoxy-phenolic blend before insulating.
- Below 121°C (250°F), blanket insulation is avoided so as to eliminate possible sagging and moisture pickup. Consideration should be given to the use of cellular glass.
- Installation of jacketing in order to provide watertight joints is emphasized. Stainless steel jacketing over insulation on stainless piping or equipment is recommended when there is any possibility of the piping or equipment reaching a metal temperature above 650°C (1200°F). Aluminum jacketing can be used; however, extreme care must be exercised to avoid contact between the aluminum and stainless steel. For example, a scrap from the trim-

TABLE 1—*Recommended insulating material.*

Material	Temperature Range, °C	Application
Mineral fibers	atmospheric temperature to 1038	piping
	93 to 1038	vessels
Mineral fibers and binder	atmospheric temperature to 816	piping
	93 to 816	vessels
Expanded perlite	atmospheric temperature to 816	piping
	93 to 816	vessels
Hydrous calcium silicate	to 816	piping
	to 816	vessels
Diatomaceous silica	to 1100	piping
	to 1100	vessels
Mineral wool and glass fiber	atmospheric temperature to 1038	piping
	atmospheric temperature to 1038	vessels
Nonrigid glass fiber	atmospheric temperature to 177	piping
	atmospheric temperature to 177	vessels
Rigid glass fiber	atmospheric temperature to 232	piping
Cellular glass	−268 to 250	piping
	−268 to 250	vessels
Ceramic fiber	to 1260	cements, castable blocks and blanket
Insulation finishing	atmospheric temperature to 1038	miscellaneous finish
Insulating cement	atmospheric temperature to 982	built-up insulation

ming of the jacketing coming in contact with stainless steel could melt and cause liquid metal corrosion resulting in rapid premature failure of the stainless equipment.

Accessories—The accessories are as follows:

- Materials used for welding studs and support clips are carbon steel for carbon and alloy steel, and American Iron and Steel Institute (AISI) Type 304 (Unified Numbering System [UNS] S30400) stainless steel for austenitic stainless steel vessels. All support rings are carbon steel if not welded directly to alloy vessels. Welding and heat treatment of attachments meet the fabrication specifications for the vessel.
- Tie wires, bands, band seals, wire mesh, and self-tapping screws for attaching insulation and aluminum sheets to equipment are 18 Cr, 8 Ni stainless steel. Bands and band seals may be aluminum for thin pipe jacketing.
- Where insulation can not be used because of the irregular surface (such as on valves, fittings, and flanges), insulating cement is used. Insulating cement exposed to the weather is waterproofed with a coat of asphalt emulsion.

Installation Practices

Surface preparation of nonaustenitic piping and equipment having operating temperatures of 10 to 121°C (50 to 250°F) involves cleaning (Steel Structure Painting Council Grade SP-5) and coating with a rust inhibiting type primer before applying insulation.

The following procedures are strictly observed:

1. Rigid type insulation is applied with staggered joints.
2. Double layer construction is used for equipment and piping in all services above 260°C (500°F).
3. In multiple layer, rigid type insulation, both longitudinal and end joints of the outer layer are staggered with respect to joints of inner layers. Each layer is separately secured. Support rings for insulation are provided on straight vertical piping runs. The maximum spacing of rings is 6 m (20 ft). Insulated flanges are not used for vertical insulation support.
4. Insulation materials must be kept dry until protective coating or jacketing is installed.

Research on Coatings for Stainless Steel

In following the current recommended practice for insulating austenitic stainless steel, our company has experienced difficulties in successfully coating stainless steel equipment before insulating. These difficulties primarily result from the poor adhesion of most coatings to stainless steel. Stainless steels form a very dense smooth oxide film at their surface that greatly reduces the amount of mechanical bonding between the steel and a coating. In addition, the oxide at the surface provides a different chemistry than a carbon steel for chemical bonding with the coating.

In 1981, we began a research program to evaluate the performance of coatings on austenitic stainless steels. It was hoped that the results of this research would allow us to improve our recommended practices for insulating stainless steel equipment.

A review of the literature indicated that silicones and epoxies are being used to coat stainless steel [*1-5*]. Based on this information and on manufacturer's recommendations, seven coatings were selected for the initial test program, including five silicones, an epoxy, and a vinyl. Coatings tested in the second phase were selected primarily on the basis of the previous results, with the addition of one new coating. Tables 2 and 3 give additional information on all the coatings that have been tested.

Test Program

The tested coatings (described in Tables 4 and 5) were applied to 7.6 by 22.9-cm (3 by 9-in.) panels of Type 304 stainless steel for all tests except the outdoor exposure. In the first phase, our in-house personnel applied the coatings over solvent cleaned panels. For the second phase, the manufacturers supplied us with coatings over solvent cleaned (only) and sandblasted (approximately 1/2-mil surface profile) surface preparation.

The test program was designed to evaluate

(1) adhesion to stainless steel,
(2) flexibility, and

TABLE 2—*Manufacturer's information, coatings for first phase of tests.*

Coating Identification Number	Coating Type	Color Used in Tests	Manufacturer's Recommended Maximum Service Temperature, °C
8233	silicone	black	425 continuous 540 excursions
8234	silicone	black	455
8235	vinyl copolymer	gray	71 continuous 82 excursions
8236	modified silicone	black	400 continuous 540 excursions
8237	copolymerized silicone	white	260 continuous 290 excursions
8238	epoxy phenolic	green	140
8239	silicone alkyd and silicone	clear	150

TABLE 3—*Manufacturer's information, coatings for second phase of tests.*

Coating Identification Number	Coating Type	Color Used in Tests	Manufacturer's Recommended Maximum Service Temperature, °C
8318	silicone	black	455
8321	vinyl copolymer	gray	71 continuous 82 excursions
8322	epoxy polyamide	black	95
8323	modified silicone	black	400 continuous 540 excursions
8324	silicone	black	425 continuous 540 excursions
8325	copolymerized silicone	white	260 continuous 290 excursions

NOTE: Coating 8233 and 8324 are the same. Coating 8234 and 8318 are the same. Coating 8235 and 8321 are the same. Coating 8236 and 8323 are the same. Coating 8237 and 8325 are the same.

TABLE 4—*Application information (Phase 1).*

Coating #	Type of Coating	Number of Coats	Film Thickness, mil dry	Comments
8233	silicone	2(CS) 2(SS)	3 to 4 2.5 to 3	Dry spraying occurred on SS and CS surfaces leaving a nonuniform film (that is, patchy coverage).
8234	silicone	2(CS) 2(SS)	4 to 5 2 to 2.5	sagging problems with first coat thinner not used with second coat
8235	vinyl copolymer	2(CS) 2(SS)	4 to 5 3.5 to 4	second coat went on heavier than first
8236	silicone	2(CS) 2(SS)	7 to 8 4 to 4.5	...
8237	silicone copolymer	2(CS) 2(SS)	5 to 7 2.5	...
8238	epoxy-phenolic	1(CS) 1(SS)	3 to 4 3 to 4	only one coat applied
8239	silicone alkyd	2(CS) 2(SS)	1 to 1.5 1 to 1.5	unable to achieve recommended film thickness; coating remained tacky after five days; product slow to cure

NOTES: Material was applied over the following type surfaces: (1) carbon steel (CS)—sandblasted to white metal (SSPC-SP5 or NACE #1) and (2) stainless steel (SS)—solvent wiped (SSPC-SP1). Conventional pressure pot spray equipment was used. 1 mil = 10×10^{-5} m.

TABLE 5—*Application information (Phase 2).*

Coating Identification Number	Coating Type	Film Thickness, mil dry	Comments
8318	silicone	4 to 4.5	applied over solvent cleaned stainless steel
8318A	silicone	4 to 4.5	applied over blasted stainless steel
8321	vinyl copolymer	3 to 4.0	applied over solvent cleaned stainless steel
8321A	vinyl copolymer	3 to 4.0	applied over blasted stainless steel
8322	epoxy polyamide	3 to 4.0	applied over solvent cleaned stainless steel
8322A	epoxy polyamide	3 to 4.0	applied over blasted stainless steel
8323	modified silicone	2.0	applied over solvent cleaned stainless steel
8323A	modified silicone	2.0	applied over blasted stainless steel
8324	silicone	3 to 5.0	applied over solvent cleaned stainless steel
8324A	silicone	3 to 5.0	applied over blasted stainless steel
8325	copolymerized silicone	3 to 5.0	applied over solvent cleaned stainless steel
8325A	copolymerized silicone	3 to 5.0	applied over blasted stainless steel

NOTE: All coating systems in Phase 2 were applied by the manufacturer. 1 mil = 10×10^{-5} m.

(3) resistance to various environments (weather, salt fog, industrial gases, elevated temperatures, and thermal shock).

Adhesion of the coatings to stainless steel was evaluated using both the scratch test and the Elcometer Adhesion Test. The scratch test is performed by scribing a crosshatched ("#") pattern through the coating and firmly applying a piece of Scotch® tape. The tape is pulled off with a firm jerk, and the adhesion evaluated by the amount of coating that comes off with the tape. In the Elcometer Adhesion Test, metal disks, approximately 2 cm (0.75 in.) in diameter, are attached to the coated panel using an epoxy adhesive, which is allowed to cure for five days. An Elcometer Adhesion Tester is attached to the disk and measures the force necessary to pull the disk and attached coating off the stainless steel substrate. The strength of adhesion is measured in megapascals (MPa).

To test coating flexibility, a coated stainless steel panel is placed over a cylindrical 2.5-cm (1-in.) diameter mandrel and bent 180° within a 15-s time span. The coatings are inspected for cracking and disbonding.

For the outdoor exposure test, each coating is applied to 15.2 by 22.9-cm (6 by 9-in.) panels of carbon steel and 304 stainless steel, which were placed on outdoor racks at Harmarville, PA and Port Arthur, TX. The panels are evaluated annually.

Coated stainless steel panels are placed in a standard salt fog (ASTM Salt Spray [Fog] Testing [B 117]) cabinet at 35°C (95°F) and 100% relative humidity. The panels are evaluated after four months and after one year.

Coated stainless steel panels are placed in an environmental cabinet containing a gaseous mixture of hydrogen sulfide, carbon dioxide, nitrogen dioxide (0.67 volume % each), and air, saturated with water at 33°C (91°F). The panels are evaluated after 24 and 48 h.

Coated stainless steel panels are placed in a Xenon-Weatherometer (ASTM Recommended Practice for Operating Light-Exposure Apparatus (Xenon-Arc Type) With and Without Water for Exposure of Nonmetallic Materials [G 26]) at 63°C (145°F) where they are exposed to Xenon-arc light and cyclic water spray (18 min water spray, and 102 min no water spray). Panels are evaluated after 250 and 500 h.

To test for high-temperature stability, coated panels are placed in furnaces at 260 and 400°C (500 and 750°F) and evaluated after 24 h and 4 months. Coatings that remain intact after the four-month exposure are retested for adhesion with the Elcometer Adhesion Tester.

To test for ability to withstand thermal shock, coated panels are placed in a furnace at 400°C (750°F) for a minimum of 3 h and then immersed in room temperature water. This cycle is repeated eight times. The panels are evaluated after each cycle.

Results

Results of the adhesion, flexibility, outdoor exposure, salt fog cabinet, environmental cabinet, and Weatherometer tests are summarized in Tables 6 and 7. Tables 8 and 9 summarize the results of the elevated temperature and thermal shock tests.

Good adhesion of a coating to stainless steel has generally been considered a difficult goal to achieve because of the smooth oxide layer at the surface of the metal. Our results would tend to confirm this, with most coatings having only fair adhesion to stainless steel. The exceptions were the epoxy polyamide (#8322) and vinyl copolymer (#8235 and #8321), which both demonstrated good adhesion. Adhesion of the coatings was essentially the same for solvent cleaned (only) and sandblasted panels, indicating that the additional effort of sandblasting is unjustified. Flexibility, which is less dependent on the substrate, was good for a number of the coatings tested.

Despite only fair adhesion, most coatings performed adequately in the outdoor exposure, salt fog cabinet, environmental cabinet, and Weatherometer tests. The copolymerized silicone (#8232 and #8325), tended to blister in tests that constantly wetted the panel surface. The silicone alkyd, #8239, had little

TABLE 6—*Results of tests at ambient and near ambient temperatures (Phase 1).*

Coating Number	Coating Type	Scratch Adhesion	Elcometer Adhesion, MPa (psi)	Flexibility	Outdoor[a] Exposure	Salt Fog[b] Cabinet	Environmental Cabinet	Weatherometer
8233	silicone	fair	1.25 (180)	good	fair (carbon steel was rusted)	poor (rust at edges)	good	good
8234	silicone	fair	0.95 (135)	good	good	fair	good (slight color change)	good
8235	vinyl	good	4.60 (665)	good	good	good	good (slight color change)	fair (slight color change)
8236	modified silicone	fair	2.25 (325)	good	good	good	good (slight color change)	fair (color change)
8237	copolymerized silicone	fair	1.40 (200)	poor (cracks)	fair	fair (blisters)	fair (blisters)	fair (color change)
8238	epoxy phenolic	fair	1.65 (240)	poor (cracks)	good	good	good (color change)	fair (slight color change)
8239	silicone alkyd + silicone	poor	0.50 (75)	good	poor (scratches and holes)	poor (peeling)	poor (alligatoring and cracks)	fair (color change)

[a] After one year exposure.
[b] Based on one year evaluation.

TABLE 7—*Results of tests at ambient and near ambient temperatures (Phase 2).*

Coating Number	Coating Type	Scratch Adhesion	Elcometer Adhesion, MPa (psi)	Flexibility	Outdoor[a] Exposure	Salt Fog[b] Cabinet	Environmental Cabinet	Weatherometer
8318	silicone	good	1.40 (200)	fair-good	N/A	excellent	excellent	excellent
8318A	silicone	good	1.25 (183)	fair-good	N/A	excellent	excellent	excellent
8321	vinyl copolymer	excellent	2.80 (400)	excellent	N/A	excellent	fair-good	good
8321A	vinyl copolymer	excellent	2.60 (375)	excellent	N/A	fair	fair-good	good
8322	epoxy polyamide	good	5.45 (783)	excellent	N/A	excellent	excellent	excellent
8322A	epoxy polyamide	good	4.75 (638)	excellent	N/A	excellent	excellent	excellent
8323	modified silicone	good	1.45 (208)	excellent	N/A	fair	fair	poor
8323A	modified silicone	good	1.40 (200)	excellent	N/A	fair	poor	poor
8324	silicone	excellent	1.40 (200)	excellent	N/A	poor	fair	fair
8324A	silicone	excellent	1.45 (208)	excellent	N/A	poor	fair	fair
8325	copoly-merized silicone	good	1.60 (225)	fair	N/A	fair	fair	excellent
8325A	copoly-merized silicone	good	1.50 (216)	fair	N/A	fair	fair	excellent

[a] N/A not available.
[b] Based on two months evaluation.

TABLE 8—*Results of tests at elevated temperatures (Phase 1).*

Coating Number	Coating Type	260°C (500°F)			400°C (750°F)			Thermal Shock
		24 h	4 Months	Elcometer Adhesion,[a] MPa (psi)	24 h	4 Months	Elcometer Adhesion,[b] MPa (psi)	
8233	silicone	good	good	1.40 (200)	good	good	2.30 (330)	good
8234	silicone	good	good	1.60 (230)	good	good (coating lost its gloss)	1.65 (240)	good
8235	vinyl copolymer	fair (color change)	poor (rubs off)	...	poor (gone)	...	...	poor
8236	modified silicone	good	poor (peeled off)	...	poor (peeled off)	...	...	poor
8237	copoly-merized silicone	fair (color change)	fair (blisters and color change)	1.10 (160)	poor (blisters)	poor (flaked off)		good
8238	epoxy phenolic	poor (flaked off)	...	...	poor (gone)	...	...	poor
8239	silicone alkyd + silicone	poor (burned off)	...	...	poor (burned off)	...	...	poor

[a] Room temperature test after 4 months exposure at 260°C (500°F).
[b] Room temperature test after 4 months exposure at 400°C (750°F).

TABLE 9—*Results of tests at elevated temperatures (Phase 2).*

Coating Number	Coating Type	260°C (500°F)		400°C (750°F)		Thermal Shock
		24 h	2 Months	24 h	2 Months	
8318	silicone	good	good	good	good	good
8318A	silicone	good	good	good	good	good
8321	vinyl copolymer	fair	fair	fair	poor	poor
8321A	vinyl copolymer	fair	fair	poor	poor	poor
8322	epoxy polyamide	fair	fail	poor	fail	poor
8322A	epoxy polyamide	fair	fail	poor	fail	poor
8323	modified silicone	good	good	fair	fair	good
8323A	modified silicone	good	good	fair	fair	good
8324	silicone	good	good	good	good	good
8324A	silicone	good	good	good	good	good
8325	copoly-merized silicone	good	good	good	good	good
8325A	copoly-merized silicone	good	good	good	good	good

resistance to chemical or mechanical damage. One of the silicones, coating #8233, had difficulties in several tests, primarily because of patchy coverage.

In the elevated temperature and thermal shock tests, only the straight silicones (#8233, #8234, #8318, and #8324) performed adequately in both tests. Adhesion of coatings (#8233 and #8234) to the stainless steel substrate was actually improved after four months at 260°C (500°F), probably because of additional curing. The copolymerized silicone (#8237) performed well for short periods of time at 260°C (500°F), but blistered and lost adhesion after four months exposure. Surprisingly, it survived the thermal shock test despite 400°C (750°F) temperatures that exceeded the manufacturers maximum recommended temperature of 290°C (550°F). Coating #8236, a modified silicone, was inconsistent in its performance. As expected, coatings #8235, #8238, #8322, and #8239 flaked or burned off in the elevated temperature tests.

Conclusions

Results of testing to date indicate the following:

1. Little or no difference was noticed in the level of performance of coatings applied over solvent cleaned stainless steel and coatings applied over blasted stainless steel.

2. Most coatings tested perform adequately on stainless steel at ambient and near-ambient temperatures despite only fair adhesion. However, the performance of the epoxy polyamide and the vinyl copolymer were clearly superior to the others.

3. At elevated temperatures, straight silicones appear to give the best performance.

Changes in Current Insulating Practice for Stainless Steels

In the past, some of our field personnel have been reluctant to apply coatings over anything less than a blasted surface. We are now initiating field tests of materials applied over solvent cleaned (only) stainless steel surfaces because of the results of this research program. In addition, we have identified new materials and suppliers for our current recommended coating systems.

References

[*1*] Dillon, C. P. and Associates, *Stress Corrosion Cracking of Stainless Steels and Nickel-Base Alloys*, Materials Technology Institute of the Chemical Process Industries, Inc., 1979, pp. 57–62, 104, 143–144.

[*2*] Mersberg, A. R. and Wee, F. W., *Materials Performance*, Vol. 19, No. 12, Columbus, OH, Dec. 1980, p. 13.

[*3*] Willhelm, A. C., "Protective Coatings to Resist Salt Corrosion and Heating to 650°F," *Proceedings of the Air Force Materials Lab 50th Anniversary Technical Conference on Corrosion of Military and Aerospace Equipment*, Denver, CO, 1967, p. 1581.

[*4*] Pilla, G. J. and DeLuccia, J. J., *Metals Progress*, Vol. 117, No. 6, June 1980, p. 57.

[*5*] Vegdahl, E. J., Damin, D. G., and Sumbry, L. C., "Eclectic Material Problems in the Petrochemical Industry," Preprint No. 17, Meeting of the National Association of Corrosion Engineers, Anaheim, CA, 18–22 April 1983.

Charles T. Mettam[1]

Designing to Prevent Corrosion of Metals Under Insulation

REFERENCE: Mettam, C. T., **"Designing to Prevent Corrosion of Metals Under Insulation,"** *Corrosion of Metals Under Thermal Insulation, ASTM STP 880*, W. I. Pollock and J. M. Barnhart, Eds., American Society for Testing and Materials, Philadelphia, 1985, pp. 178–187.

ABSTRACT: The paper concentrates on the corrosion of carbon steel under insulation. Austenitic steel corrosion is also mentioned. Hot and cold insulation materials are discussed with the main consideration place on "cold" problems. The current solution of the problem is to paint all carbon steel that is to be insulated and operating between −1 and 121°C. Insulation material and thickness is then selected. A moisture barrier or a vapor barrier is added to protect the insulation. Additional deterents added are vapor stops and contraction/expansion joints. A European versus an American design is examined as a conclusion. An Appendix, including protective coatings used, and drawings for additional clarification are included.

KEY WORDS: corrosion, insulation, permeability, urethane, coatings (under insulation)

For many years the only "corrosion under insulation" problem discussed was stainless steel corrosion. In an effort to solve the problem a Military Specification Insulation Materials, Thermal, with Special Corrosion and Chloride Requirements (MIL-24244) was followed that among other things limited the amount of chlorides in insulation to 600 ppm. Using the specification did not eliminate the stainless steel corrosion problem because it was then found that chlorides from the atmosphere also contributed to the corrosion. An immediate solution was to apply an inexpensive coating to austenitic steel operating between −1 and 121°C before insulating.

Lately, severe carbon steel corrosion under insulation or fireproofing or both has become a main concern. Studies were conducted to investigate the

[1]Senior piping engineer, Lummus Crest Inc., 1515 Broad St., Bloomfield, NJ 07090.

problem. Many companies were unaware of their insulated carbon steel corrosion problems. Included in our investigation was an ethylene plant constructed about 5 to 10 years ago. A "cold" piping system was examined. The insulation used was urethane with an inadequate vapor barrier. The moisture condition caused by "ice melt" was so bad that it was difficult to continue our task and stay dry. The observed failure of the insulation system resulted in large-scale corrosion of the structural steel supporting the vessels and pipe. Replacement of the steel at great expense and downtime to the owner was the inevitable outcome. Another large company recently interviewed revised their specifications and now requires two coats of coal tar epoxy at 200 μm/coat under insulation. The change is indicative of a solution to a definite corrosion problem. Corrosion under insulation is indeed a very serious condition as is evidenced by the number of people attending this meeting. Extensive repair must be made to rectify the problem, especially in cold service. The repairs are very costly. They are difficult to make while systems are on stream, and in some cases unit shutdowns are necessary before repairs can be made.

Design for Corrosion Under Insulation

Our solution to the corrosion problem is serious consideration during the design phase of a project. The first step is the painting specification (see Appendix). Epoxy coating was selected as the primary line of defense under the insulation. The coating works very well as the generic primer normally recommended by fireproofing suppliers. As shown in a commercial blast (Appendix) with 100-μm, dry film thickness is used. At higher temperatures the coating will change color and may even fail, however, any electrolytes present will not normally cause corrosion. At these elevated temperatures the epoxy specified is hard and very resistant to acidic or basic attack. It will accept spray on insulation as well as block and blanket type insulation.

The next step in our design is proper insulation selection. We normally specify calcium silicate, fiberglass, mineral wool, perlite, or urethane for hot insulation. Cellular glass and urethane are specified for cold insulation. Fiberglass is one type of insulation that is not recommended for cold service. Manufacturers' literature tabulates fiber glass thickness based on temperature and relative humidity; however, our experience has been that manufacturers are unable to show us successful fiber glass installation for cold service. A very recent client experienced actual failures of fiber glass cold insulation, which reinforces our stand.

Moisture prevention is our third major consideration. We achieve success by utilizing metal jacketing with a heat sealed moisture barrier over hot insulation. Bands are placed in locations so that overlaps and wide openings do not permit entry of water. To prevent the insulation from getting wet during construction, it is specified that no insulation be left uncovered after working hours. Compliance with this is very hard to obtain and the activity should be monitored. On a job a few years ago, the heating up of an insulated line was witnessed. Water poured from the jacketing to such an extent that it appeared that a weld had failed. It turned out that the insulation was soaked during recent rainstorms, and the water was being expelled in the warming process.

We impose special requirements on cold insulation handling. Every effort is made to ensure that the vapor barrier is not mechanically damaged during installation. Injury to the barrier permits warm circulating air and subsequent moisture to reach the steel.

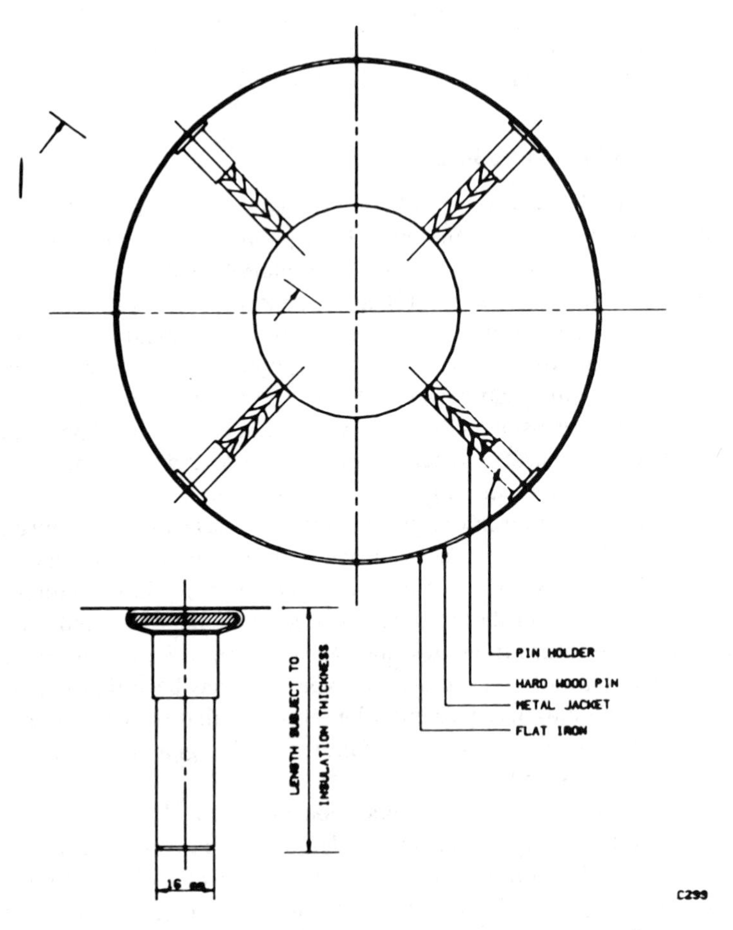

FIG. 1—*Cold insulation for foamed in situ spacing construction.*

During a recent overseas project a European company's cold insulation system was reviewed [*1*]. A facility insulated by them having two units similar to the one we were constructing was visited. One unit was operating and the other was down for a "turn around." Signs of icing on the unit in service were not present. Plant personnel had stripped the "foamed in-place urethane" from some of the lines and equipment on the "down" unit. We saw no signs of corrosion, in fact, some of the painted pipe identifications were still legible after ten years.

Figures 1 through 3 are part of the company's design that went into the

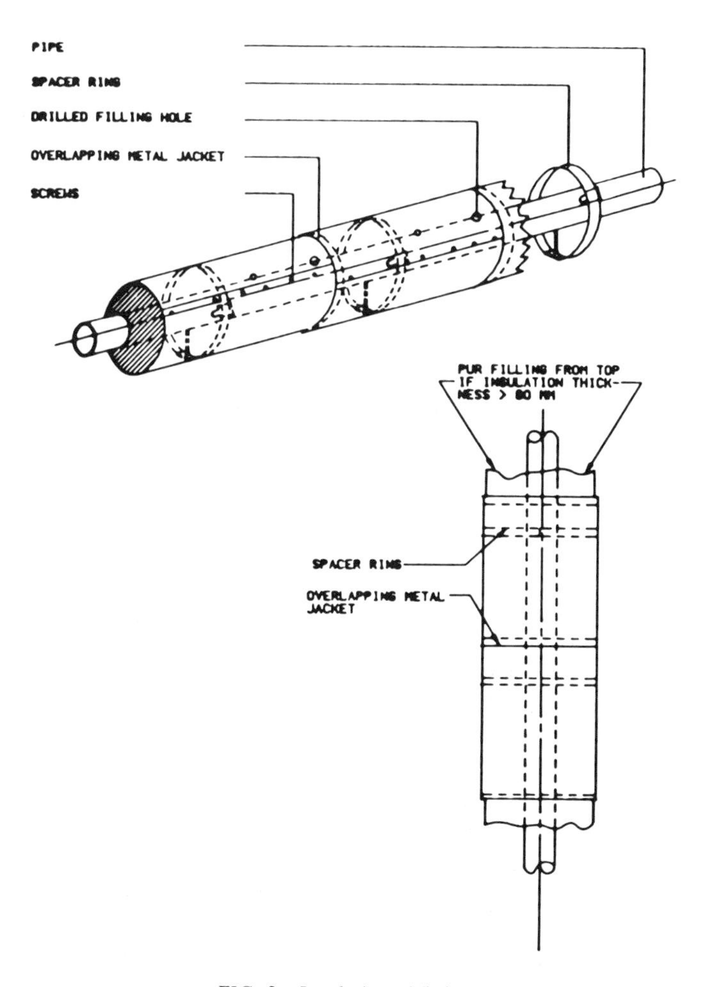

FIG. 2—*Insulation of fitting.*

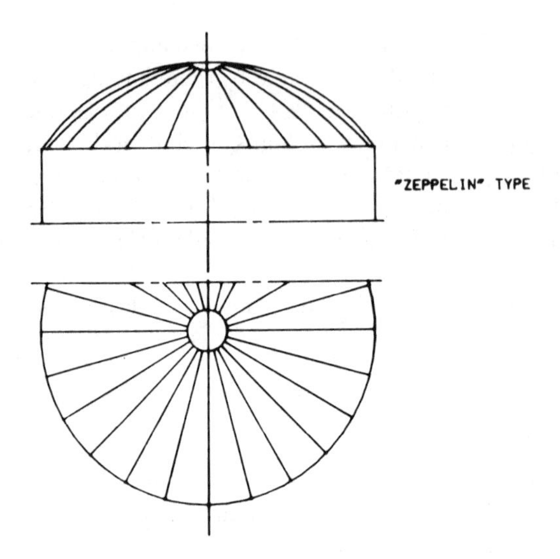

FIG. 3—*Form of vesses head.*

insulation project. The specifications were more than adequate. The main design ingredients the European contractor used for prevention of corrosion are metal jacketing, sealing tape, plastic grommets, sealer, and a government regulation, "Insulation Works—Protection Against Corrosion for Cold and Hot Insulation at Industrial Plants." Their methods are summarized as follows:

1. The initial step is painting the steel.
2. Complete metal jacketing is then used for the insulation of horizontal equipment, as well as for piping, flanges, valves, and so forth. The object to be insulated is completely covered with metal jacketing and spacers. A telescopic metal jacketing can be used for the insulation of vertical equipment and for vertical piping. Rings of flat iron with spacers are placed at a distance of 1 m maximum. Urethane blocks are used as spacers if rings cannot be installed. The jacketing is cut to size and installed with an overlap of 50 mm on both horizontal and vertical seams. The sheets are fixed with self-tapping screws at a maximum of 100 mm distance. The jacketing follows the outline of the equipment. Curved heads are covered using the Zepplin shape as shown in Fig. 3. In order to attain a higher stiffness the sections of sheet metal are hemmed. Supports, skirts, and so forth in direct contact with equipment and penetrating the insulation are insulated for a length equal to five times the insulation thickness with a minimum of 300 mm.

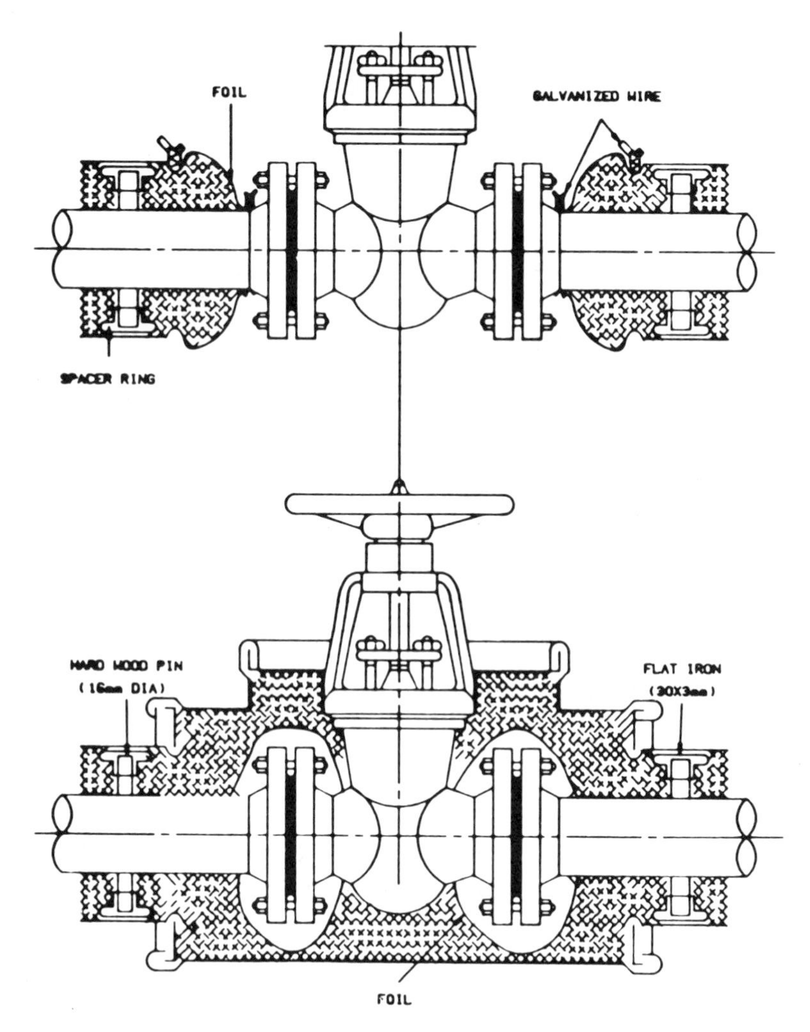

FIG. 4—*Insulation of valves.*

3. Foam is then pumped in through openings drilled into the metal jacketing. The openings are fitted with grommets after pumping.

4. If parts of the equipment, such as manholes or nozzles, cannot be insulated at the same time as the equipment, jacketing must be installed very close to these parts and banded tight. The jacketing is cut back when the parts are insulated. This is necessary to make a good foam connection and to avoid thermal leaks.

5. Finally all joints, laps, and crimps of the metal jacket are covered with

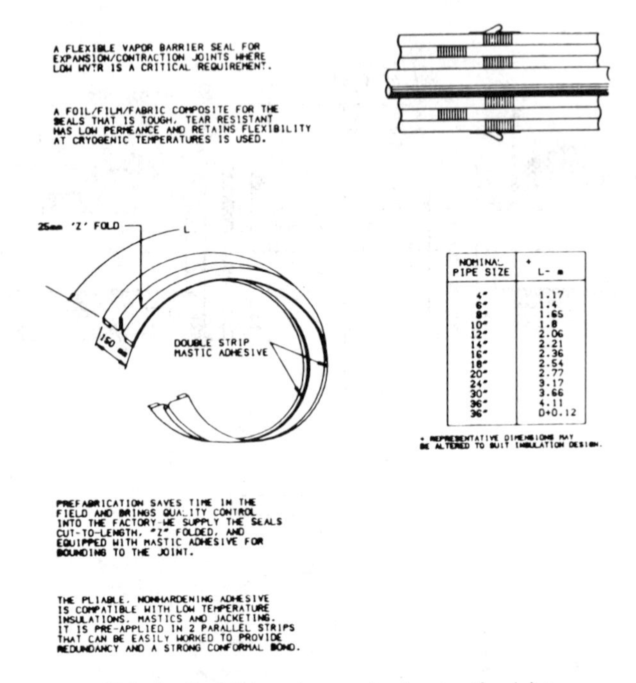

NOMINAL PIPE SIZE	* L- m
4"	1.17
6"	1.4
8"	1.65
10"	1.8
12"	2.06
14"	2.21
16"	2.36
18"	2.54
20"	2.77
24"	3.17
30"	3.66
36"	4.11
36"	D+0.12

* REPRESENTATIVE DIMENSIONS MAY BE ALTERED TO SUIT INSULATION DESIGN.

FIG. 5—*Prefabricated expansion/contraction joint.*

sealing tapes. The tape must be permanently elastic and weather resistant. As stated previously, our practice is to specify cellular glass or urethane for cold insulation. In our design one of the most important items other than insulation thickness is the vapor barrier used. Normally used is a product with a low water vapor permeability. Most material can be sprayed or trowelled on. With either of these methods the thickness of the material can vary considerably, unless it is closely monitored. To avoid thickness inconsistency, we specify the product to be supplied in sheets of standard thickness. Also specified is the adhesive to be used for bonding the sheets together. Pipe fittings have not yet been developed for sheets so a sprayed or trowelled on vapor barrier is specified. Valves are handled slightly different (Fig. 4). The use of prefabricated expansion/contraction joint seals [*2*] is a necessity in designing against failures leading to corrosion (Fig. 5). Since our best design does not prevent human error or acts of God, termination seals [*3*] are used (Fig. 6) to prevent a failure from propagating into the system. Our design is monitored periodically by our engineering staff with visits to projects after they have been in operation a while.

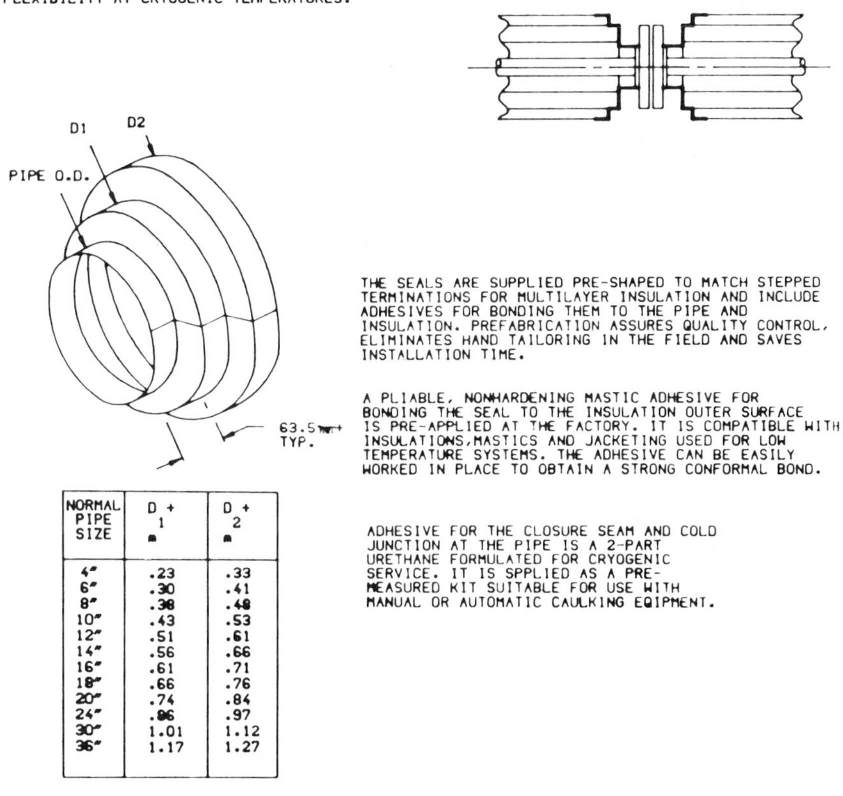

NORMAL PIPE SIZE	D + 1 m	D + 2 m
4"	.23	.33
6"	.30	.41
8"	.38	.48
10"	.43	.53
12"	.51	.61
14"	.56	.66
16"	.61	.71
18"	.66	.76
20"	.74	.84
24"	.86	.97
30"	1.01	1.12
36"	1.17	1.27

FIG. 6—*Prefabricated termination seal.*

A representative recently inspected a process plant that we built approximately six years ago. Some of the design features [*4*] of the cold insulation on that project are shown in Fig. 7. The insulation, piping, and related areas were in excellent condition. The only places that showed any signs of failure were where insulation was removed in order to get at a valve or flange. The maintenance personnel failed to reinsulate correctly, and ice is now forming in these areas.

The selection of a good contractor, careful installation inspection, and adherence to our design specifications insure an excellent insulation job.

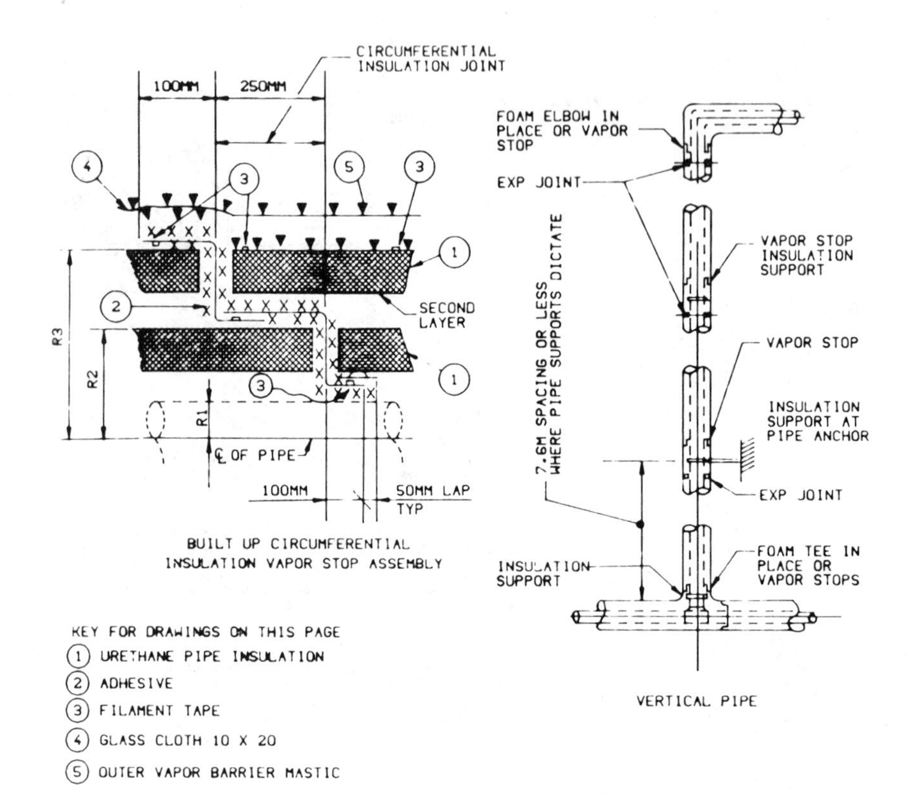

FIG. 7—*Design features of the cold insulation.*

APPENDIX

Surface Preparation and Coating System for Insulated and Fireproofed Carbon and Low Alloy Steel

Limitation

1. Where metal surface temperature is between −1° and 121°.
2. Vessels-tanks, towers, drums, fireproofed skirts, fireproofed heads, fireproofed structural steel, fired heaters, exchangers and similar equipment, pipe hangers, and piping.
3. Priming is to be done in the shop unless otherwise specified.

TABLE 1—*Coating system: epoxy system/epoxy phenolic.*

Manufacturer	Primer, Minimum Dry Film Thickness, μm
A	A-1 (epoxy phenolic) 100
B	B-1 (amine epoxy) 100
C	C-1 (polyamide epoxy) 100

NOTES: Paint manufacturer's recommendation for painting application, such as mixing of paint, application equipment, and so forth, shall be ridgidly followed.

Surface Preparation

1. Before blasting, grind smooth-sharp edges, rough welds, steel silvers, and so forth, remove all oil, and weld spatter and flux.
2. Blasting media shall be a 16 to 40 mesh nonmetallic abrasive, a Society of Automotive Engineers (SAE) GL-40 steel grit, a S-230 steel shot or equivalent abrasive that will give a minimum surface profile of 25 μm deep and maximum of 65 μm with rogue peaks not exceeding 75 μm. Blasting media shall be free of all oil and moisture.
3. Blast clean all steel surfaces and welds to be coated to SSPC-SP6.
4. No blasted surfaces shall be allowed to remain uncoated overnight in humid areas. All paints shall be applied to a surface free from moisture, oil, dust, grit, or any other contaminants and discoloration. The initial blast quality shall be maintained immediately before painting.

References

[1] Rheinhold and Mahla, "Cold Insulation for Equipment and Piping with Polyurethane (PUR) Foamed in Situ," Project Engineering Specification, Mannheim, West Germany, Aug. 1977.

[2] Sheldahl, "Flexible Vapor Barrier Seal for Insulated Terminations," Northfield, MN, 1982.

[3] Sheldahl, "Flexible Vapor Barrier Seal for Insulation Expansion/Contraction Joints," Northfield, MN, 1982.

[4] Chicago Bridge and Iron Co., "Specification for Multi-Layer Polyurethane Pipe Insulation," Oak Park, IL, June 1976.

James A. Richardson[1] *and Trevor Fitzsimmons*[1]

Use of Aluminum Foil for Prevention of Stress Corrosion Cracking of Austenitic Stainless Steel Under Thermal Insulation

---

REFERENCE: Richardson, J. A. and Fitzsimmons, T., **"Use of Aluminum Foil for Prevention of Stress Corrosion Cracking of Austenitic Stainless Steel Under Thermal Insulation,"** *Corrosion of Metals Under Thermal Insulation, ASTM STP 880*, W. I. Pollock and J. M. Barnhart, Eds., American Society for Testing and Materials, Philadelphia, 1985, pp. 188–198.

ABSTRACT: For many years, it has been preferred practice within Imperial Chemical Industries (ICI) to use aluminum foil on austenitic stainless steel surfaces operating at temperatures within the range 60 to 500°C. The foil protects in two ways. It presents a physical barrier to chloride-containing fluids migrating through lagging materials towards hot stainless steel surfaces. It also cathodically protects stainless steel in the event of flooding of the lagging system, thereby preventing initiation of pitting and stress corrosion cracking. ICI's experience with aluminum foil is summarized. Laboratory data are presented that confirm the galvanic protection afforded by the foil, and the efficiency of the foil in preventing chloride stress corrosion cracking relative to a number of coating systems specified for the same purpose.

KEY WORDS: aluminum foil, stress corrosion, austenitic stainless steel, thermal insulation, cathodic protection, silicone-alkyd coatings, aluminum-rich silicone coatings, zinc-rich coatings

External stress corrosion cracking of austenitic stainless steel is a long-standing problem in the process industries. Its phenomenology has been discussed admirably in numerous previous publications and in other papers in this STP, and it is not necessary to describe it in detail in this paper. For the problem to occur, three basic conditions must be met:

[1] Senior corrosion engineer and materials engineer, respectively, Imperial Chemical Industries PLC, Engineering Department—North East Group, P.O. Box 6, Billingham, Cleveland TS23 1LD, United Kingdom.

1. The stainless steel surfaces must operate continuously, or intermittently, at temperatures > approximately 60°C. The few significant failures we have experienced in Imperial Chemical Industries (ICI) over the past 10 to 15 years have been on surfaces operating at < approximately 200°C.
2. There must be soluble chlorides present. This topic has been discussed in detail elsewhere,[2] but suffice it to say that virtually all insulating materials, perhaps with the exception of cellular glass, contain some soluble chloride, and that chloride is a common contaminant in air, rainwater, most process waters, and other process fluids that can enter lagging systems.
3. Water must enter the lagging system. In terms of preventing stress corrosion cracking it is convenient to distinguish two "degress" of water ingress:

    a. Migration—Weatherproofing/vapor sealing systems are rarely 100% efficient. At best the lagging system will still be able to breathe allowing access of atmospheric moisture, and in addition there may be minor leaks in the seals allowing ingress of small quantities of rainwater, and so forth. The extent to which such water will migrate towards, and wet stainless steel surfaces will depend upon the "wicking" properties (if any) of the insulation material, and the process temperature in relation to the dew point of the lagging "atmosphere." Given that process temperatures may vary and that there are diurnal and seasonal variations in temperatures and humidity, there is obviously scope for alternate wetting/drying phenomena to concentrate any chloride present at the metal surface.
    b. Flooding—Where weatherproofing or vapor sealing systems have been poorly designed or executed or both, or have deteriorated significantly after prolonged periods of service, then ingress of substantial quantities of water is possible. Common sources of aqueous fluids include, apart from rainfall, wash waters, condensate from steam tracing joint leaks, quench system waters, and process fluids leaking from joints. Permanently flooded lagging systems offer little, if any, thermal insulation and are thus likely to attract attention. However, intermittent flooding may well go undetected, or at least be tolerated, together with the attendant mechanisms for carrying soluble chlorides to the hot metal surface where they can concentrate in a series of wetting/drying cycles.

Accepting these basic conditions for cracking, it follows that the first and obvious preventative measure is to keep water out of lagging systems. Lagging systems need their share of "good engineering practice," which means appropriate levels of design, inspection, and maintenance. This topic is covered in other papers in this STP and will not be pursued further here, but its crucial importance is acknowledged. However, most process plant operators seek to provide some additional protection against stress corrosion cracking, and ICI is no exception.

[2]Richardson, J. A., in this publication, pp. 42-59.

Development of a Cracking Prevention Policy

When the problem was first addressed within ICI approximately 15 years ago, the defined policy objective was to identify a protection system that would prevent stress corrosion cracking of stainless steel surfaces operating at temperatures $>60°C$ resulting from "migration" of water (as defined above) and occasional transient flooding of the lagging system. Three basic approaches to the problem were identified.

Control of Insulation Material Composition

This approach considered the possible specification of low-chloride or inhibited lagging materials or both in the hope of controlling the corrosivity of aqueous extracts concentrating on hot stainless steel surfaces. It was concluded that this approach was not sufficient to meet the policy objective. It was recognized that the specification of materials with, for example, controlled initial $Na + SiO_2/Cl + F$ ratios [*1*] could certainly reduce the risk of cracking. However, progressive ingress of chloride via mechanisms of the type discussed above would inevitably alter adversely the ratio and periodic flooding if the lagging might well remove the soluble inhibitor. In any event, there was a reluctance to carry the additional costs of over-specifying with respect to national standards, and in 1970, for example, the appropriate British Standard (BS) for Thermal Insulating Materials (BS 3958, Part 2) allowed up to "approximately 550-ppm chloride" in preformed calcium silicate. As has been pointed out recently [*2*], the potential total chloride per unit area of stainless steel surface is of more concern in practice than the chloride content of the lagging per se.

Use of Coatings

This approach essentially involved the provision of a physical barrier between the stainless steel surface and any corrodent accumulating thereon. It was concluded that this too was insufficient to meet the policy objective. Although paints capable of operating at relatively high temperatures, for example, silicone-based formulations were available, they could only be as efficient in preventing cracking as they were free of holes and imperviousness. It would be virtually impossible to achieve a defect-free paint system under a lagging system, bearing in mind the likely damage caused during lagging application. It would be costly to achieve a near defect-free system involving multi-coats, spark, or sponge testing, and so forth. A relatively cheap, single-coat system would inevitably leave a significant risk of crack initiation at defects.

Simultaneous Provision of a Physical Barrier and Galvanic Protection

It is well known that chloride stress cracking initiation in stainless steels is potential dependent [*3,4*]. In particular, it is possible to prevent crack initiation completely with appropriate levels of cathodic protection. It was thus decided to explore the simultaneous provision of a physical barrier, to cope with "migration" of water through the insulation and sacrificial cathodic protection, to cope with protection at holidays and with occasional flooding problems. Two approaches were identified:

1. Metal-Filled Paints—Aluminum-filled paints were ruled out, because little if any galvanic benefit derives from their use because of relatively low metal loadings and aluminum oxidation characteristics. Metallic zinc-rich paints proved highly efficient at preventing cracking, but bearing in mind the relatively low melting point of zinc ($\sim 420°C$) and the attendant concerns of liquid metal embrittlement,[2] they were unlikely to find widespread use.
2. Aluminum Foil—This emerged rapidly as the leading contender among the approaches evaluated. Laboratory work established its efficiency, both as a physical barrier and in the provision of galvanic protection. Melting and liquid metal embrittlement risks were considered significantly lower compared with zinc. As long as thin gage foil was used, that is, light, virtually self supporting, compliant, and easy to join, application problems and costs could be reduced to a minimum.

The use of aluminum foil under lagging systems to prevent external stress corrosion cracking of austenitic stainless steel surfaces operating between 60 and 500°C was first introduced into ICI lagging specifications approximately 15 years ago. It is the preferred preventative system, regardless of the lagging material, and has been used on the large majority of austenitic stainless steel vessels and piping systems constructed during that period. Where use of foil has proved impracticable or undesirable for some reason (see below), paint coatings have been used, with due recognition that, with the possible exception of metallic zinc-rich systems, they are not as efficient as foil in preventing stress crack initiation.

Coating and Aluminum Foil Efficiencies in Preventing Cracking

The relative efficiencies of coating and aluminum foil in preventing cracking have been evaluated in the laboratory on a number of occasions over approximately the past 15 years. The problem is to devise a test procedure that allows valid statistical evaluation of the relative efficiencies of various preventative systems in preventing crack initiation, while remaining relatively cheap and convenient to perform. Coiled spring specimens are particularly convenient in this respect, in providing a relatively large stressed surface area for coating and for presentation to corrodents. In our recent tests, we standardized on the use of specimens constructed from 3 mm diameter, American Iron and Steel Institute (AISI) Type (304) (Unified Numbering System [UNS] S30400) stainless steel wire, cold formed into a 10-turn helix with a 20-mm internal diameter, providing a total surface area of about 65 cm^2. Typical specimens are shown in Fig. 1.

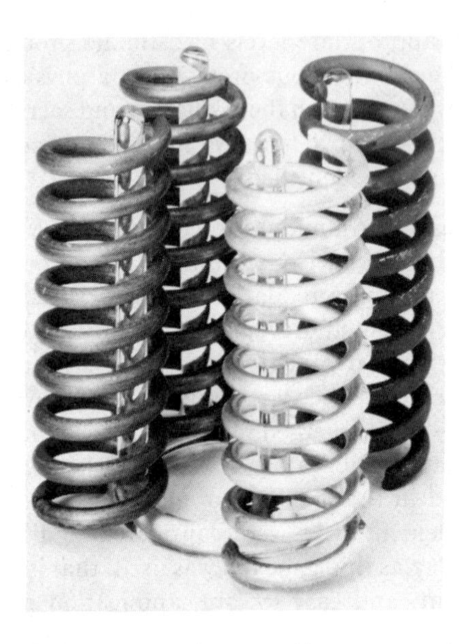

FIG. 1—*Coiled spring specimens used in stress corrosion testing.*

Specimen surfaces were degreased with acetone before brush application of single coats of paint coatings, that is, simulating site application conditions. Alternatively, specimens were wrapped externally with 46 standard wire gage (SWG) aluminum foil forming a cylinder around the coiled spring. No attempt was made to establish electrical continuity other than crimping the foil around individual turns in the spring. Specimens were subjected to full immersion testing for periods, typically seven days, in various chloride solutions in glass flasks under reflux at various boiling temperatures in the range 105 to 145°C.

At the conclusion of the tests, paint coatings were removed with appropriate solvents and specimens were dye penetrant examined using fluorescent dye under an ultraviolet (UV) lamp. The number of cracks were recorded. The existence of cracks was confirmed by metallography, and a typical crack is shown in section in Fig. 2. During some tests, natural corrosion potentials were measured using a saturated calomel electrode in a glass sidearm coupled to the test fluid by a sintered glass disk and a chloride-containing polyacrylamide gel. Some typical test data are presented in Tables 1 and 2.

The following conclusions emerged from the tests:

1. A single coat silicone-aklyd system, brush applied to a degreased stainless steel surface occasioned a 90 to 95% reduction in the incidence of cracking compared with a bare surface. Curing of the coating at 120°C before testing appeared to have little influence on protection efficiency at the levels of

FIG. 2—*Microsection of typical stress corrosion crack in coiled spring specimen.*

TABLE 1—*Incidence of stress corrosion cracking on coiled 304 spring specimens in boiling saturated sodium chloride solution at 108°C.*

Protection System	Corrosion Potential, mV/SCE[a]	Total Number of Cracks, 4 Specimens	Protection Efficiency, %
None (control)	−380	75	...
Silicone-alkyd paint, uncured	−140	8	89
Aluminum-rich silicone paint	−390	8	89
Zinc-rich epoxy paint	−720	2	97
Aluminum foil	−910	0	100

[a]Potentials recorded at the test temperature of 108°C. SCE is saturated calomel electrode.

coating integrity achieved. The apparently different protection efficiencies in the two tests quoted were considered to arise from the different corrosivities of the test liquors, rather than any variation in coating integrity between the tests.

2. An aluminum filled silicone paint proved no more efficient than a metal-free paint in preventing crack initiation, and the corrosion potentials in Table 1 confirm little or no evidence of galvanic protection.

TABLE 2—*Incidence of stress corrosion cracking on coiled 304 spring specimens in boiling saturated calcium chloride solution at 138°C.*

Protection System	Total Number of Cracks, 4 Specimens	Protection Efficiency, %
None (control)	462	
Silicone-alkyd paint, cured at 120°C for 1 h	21	95
Silicone-alkyd paint, uncured	26	94
Aluminum foil	0	100

3. The particular zinc-rich paint tested reduced markedly the incidence of cracking, and depressed the natural corrosion potential by approximately 350 mV cathodically. Indeed, two of the four specimens tested exhibited no cracks at all, and it is surprising that any cracking was experienced at such low corrosion potentials.
4. Aluminum foil proved 100% efficient at preventing cracking under the test conditions, unsurprisingly in view of the very low natural corrosion potentials achieved, involving a depression of about 530 mV relative to the control corrosion potential. At such low potentials, the stainless steel is almost certainly fully cathodically protected [*4*]. The beneficial effects of aluminum foil wrapping under Karnes test conditions have been confirmed recently [*5*].

The above data confirm the galvanic protection afforded by aluminum foil under fully flooded conditions, and its advantages over single-coat paint systems. It is important to recognize that higher integrity (and costly) paint systems, that is, spark or sponge tested multi-coat systems on blasted surfaces, would probably have performed as well as aluminum foil under test conditions. Indeed certain of the single-coat specimens achieved 100% protection in the tests quoted above, for example, one of the four springs coated with aluminum-rich silicone quoted in Table 1 exhibited no cracking after seven days. However, the purpose of the tests was to evaluate aluminum foil against some relatively cheap single-coat site application procedures, and its advantages in this respect have always been confirmed in laboratory testing.

Consideration of Fire Risks

In the case of plants handling flammable fluids, which can give rise to flammable atmospheres, there are essentially three problems to address:

1. Is there an additional risk of spread of fire associated with the presence of the foil per se?

2. Is there an additional risk of spread of fire resulting from liquid metal embrittlement of stainless steel substrates by molten aluminum?
3. Is there an increased risk of ignition of a flammable atmosphere by incendive sparks resulting from a "thermite" reaction between aluminum foil and adjacent metal surfaces?

In relation to the first question, there is some dispute as to whether aluminum foil burns in air [*6*]. However, if exposed in the event of a fire it could undoubtedly act as a source of hot, molten metal or oxide particles, which propelled by expanding air, could raise surrounding materials to their ignition temperatures [*7*]. But if the foil is entirely contained beneath a lagging system, this should not be possible unless the lagging material itself is flammable. The presence of the foil in the context described in this paper, is thus not considered to increase the risk of spread of fire.

Regarding the second issue, it is undoubtedly true that molten aluminum can cause cracking of stressed austenitic stainless steel. However, laboratory testing suggests that cracks do not initiate very readily[2] and that the risks are relatively minor compared with zinc, where the zinc/nickel reaction favors crack initiation and propagation. There is some practical evidence to support laboratory findings. For example, during one reported refinery fire, both molten zinc and aluminum had access to stainless steel piping, but embrittlement problems were restricted to those caused by zinc [*8*]. In one instance, we have experienced melting of foil because of a temperature excursion on some lagged stainless steel ducting, without any evidence of liquid metal attack. Again, given the present context involving foil beneath a lagging system, the likelihood of melting induced by external flame impingement would seem remote and the risks of spread of fire caused by embrittlement acceptably low.

Regarding the "thermite" reaction risks, it is recognized that aluminum foil if placed on rusty steel and subsequently struck by a hard object, could ignite a flammable vapor through the creation of an incendive spark [*9*]. However, there is no evidence that such sparks can be produced between aluminum and a clean stainless steel surface, and in any event, there would inevitably be some attentuation of impact energy within the lagging system where the foil is entirely contained. It has thus been concluded that the risks associated with possible aluminum foil/stainless steel surface interactions beneath a lagging system are acceptably low. However, foil is not used in locations where it could contact carbon steel surfaces in service, for example, where carbon steel backing flanges are used on stainless steel lines carrying flammable fluids. In these circumstances silicone-based paints are used.

Given the qualifications outlined above, it has been concluded that the additional fire risks associated with the use of aluminum foil are acceptably low. They must be balanced against the additional fire risks associated with external stress corrosion cracking itself, which can obviously result in the leakage of hazardous/flammable materials into the plant atmosphere.

The above comments are not to be taken as a justification for the extensive use of aluminum alloys in high fire risk areas. ICI shares the concerns of other major operators on the latter topic, and the comments above relate strictly to the context of the paper.

Practical Application and Performance of Aluminum Foil

All ICI lagging specifications require that austenitic stainless steel surfaces operating continuously or intermittently between 60 and 500°C be wrapped in 46-SWG aluminum foil. Application to pipe and vessel surfaces presents few, if any, problems, largely because of the lightness and compliance of the foil. Pipes are simply wrapped in the foil with 25-mm minimum overlaps, formed so as to shed water on vertical lines. The foil can be "molded" around flanges and fittings. On vessels, the foil is usually applied in bands in advance of the lagging system. There are no support problems, because given a relatively few anchor points, the foil is virtually self supporting. Obviously, there are no support problems on the top dome. On the sides, there are usually lagging support rings that provide anchor points. On the base, there are sprags or other means of lagging support that can be used to support the foil. Overlapped, crimped joints are executed to shed water.

In the case of steam traced lines, a double wrapping of foil if specified, one directly onto the pipe beneath the tracing and the other enclosing both tracing and pipe in order to exclude unwanted insulation from the space between them. Overlaps on vertical lines are again formed so as to be water shedding. All compressive or flanged joints in steam tracing are specified to be outside, preferably underneath, the main pipe insulation system.

Some relative typical costs of applying foil and single paint coatings are as follows:

aluminum foil, £3.30/m^2
aluminum-filled silicone, £5.70/m^2
silicone-alkyd, £5.00/m^2

The attractions of foil are self evident. Foil has the additional advantage that its application by lagging trades is accepted, whereas the use of coatings requires the involvement of additional trades in the lagging process, with consequent additional timing/management problems.

During the approximate 15-year period since aluminum foil was first specified, there have been no failures of austenitic stainless steel surfaces caused by external stress corrosion cracking where the foil had been applied to specification. Admittedly, a relatively small sample of the total lagged surface area has been available for inspection during that period, but a small number of vessels and piping systems have had their laggings removed after periods of service up to 10 years without any evidence of cracking. There have been a few instances of failure of stream traced lines during this period where the protection system failed to meet specification for one or more of the following reasons:

1. A single wrap of excessively thick, rigid material (thin sheet) had been used, instead of a double wrap of foil. Rigid material tends to stand away from pipe surface, denying intimate contact between the aluminum and the pipe surface, and making overlaps difficult.
2. Flanged joints in the steam tracing had been located so as to make weatherproofing of the adjacent lagging system virtually impossible, and so as to leak into the lagging system.
3. Steam lances had clearly been pushed into the lagging to assist in clearing blocked lines, with consequent excessive damage to the lagging system.

In all cases, there was clear evidence that significant quantities of condensate had evaporated to dryness on pipe surfaces isolated from contact with aluminum in the "cavities" between the excessively rigid sheet and the pipe surfaces. Under these circumstances, galvanic protection cannot operate.

In the latter context it is appropriate to comment on corrosion of the aluminum foil, which is <5-mil thick. Obviously, if such thin foil is exposed to flooded lagging for prolonged periods then significant corrosion is bound to occur. Experimental work reported here and elsewhere [5] indicates that galvanic protection is achieved even with a significant degree of perforation of the foil. For example, at the conclusion of the tests in saturated calcium chloride solution reported in Table 2, the foil was extensively perforated as a result of pitting corrosion. However if the gaps in the foil achieve dimensions such that aqueous extracts can concentrate on austenitic stainless steel surfaces in isolation from the foil, then galvanic protection cannot operate. Corrosion of this severity had occurred in some of the steam tracing failures cited above.

In this context it is important to recall the initial primary objective in developing the aluminum foil system, that is, that it should prevent stress corrosion cracking arising from "migration" of water and occasional transient flooding of the lagging system. None of the available stress corrosion cracking prevention systems, aluminum foil, coatings, inhibited laggings, for example, can provide unlimited protection during prolonged exposure to flooded insulation systems. This automatically imposes a primary requirement for prevention that adequate levels of weatherproofing and vapor sealing be maintained throughout a plant's life.

Finally, the few significant failures caused by external stress corrosion cracking we have experienced in ICI over the past 10 to 15 years have been on surfaces either without any protection at all or protected by coatings. In the latter category, the most significant failure involved a number of columns operating in the temperature range of 110 to 180°C, which had been coated with a zinc oxide filled silicone-alkyd paint before lagging. After approximately 10 years service at a coastal site, the Type 304 columns were found to have suffered extensive stress corrosion cracking. Cracks had presumably initiated at holidays in the coating, but had propagated extensively beneath the apparently intact coating. Indeed, the coating had to be removed by surface grinding to facilitate crack detection.

Taken overall, our experience tends to confirm the relative efficiencies of the various protection systems suggested by the laboratory work reported in Tables 1 and 2, and to reinforce our view that aluminum foil is the most efficient, cost effective available system for preventing initiation of stress corrosion cracking.

Conclusions

1. Thin, 46-SWG aluminum foil can be applied to austenitic stainless steel surfaces beneath lagging systems to prevent chloride-induced stress corrosion cracking.
2. The foil acts as a physical barrier to the "migration" of small quantities of aggressive fluid towards stainless steel surfaces, and provides galvanic protection in "flooded" lagging systems, preventing initiation of and pitting and stress corrosion cracking.
3. Laboratory tests confirm that aluminum foil is more efficient than single-coat paint systems in reducing the risk of stress corrosion crack initiation.
4. Zinc-rich paints are also highly efficient at preventing stress corrosion crack initiation, but there are more significant attendant concerns about liquid metal embrittlement than in the case of aluminum foil.

Acknowledgments

A number of members of ICI's Materials Group have contributed to the development of an aluminum foil policy over the past 10 to 15 years. In particular the authors would like to acknowledge J. G. Hines and M. E. D. Turner, much of whose distilled wisdom is presented in this paper.

References

[*1*] *U.S. Atomic Energy Commission Regulatory Guide 1.36*, "Non-metallic Thermal Insulation for Austenitic Stainless Steel."
[*2*] Nicholson, J. D., *Bulletin of the Institution of Corrosion Science and Technology*, Vol. 19, No. 5, Oct. 1981, pp. 2-5.
[*3*] Wranglen, G., *An Introduction to Corrosion and Protection of Metals*, Institute for Metallskydd, Stockholm, Sweden, 1972, pp. 117-118.
[*4*] Andresen, P. A. and Duquette, D. J., *Corrosion*, Vol. 36, No. 8, Aug. 1980, pp. 409-415.
[*5*] Gillett, J. and Johnson, K. A., presented at *Conference on Corrosion Under Lagging*, Institution of Corrosion Science and Technology, Nov. 1980, published by Fibreglass Ltd.
[*6*] West, E. G., *The Metallurgist and Materials Technologist*, Vol. 14, No. 9, Sept. 1982, pp. 395-398.
[*7*] Ruskin, A. M., *The Metallurgist and Materials Technologist*, Vol. 15, No. 7, July 1983, p. 358.
[*8*] Cantwell, J. E. and Bryant, R. E., *Hydrocarbon Processing*, May 1973, pp. 114-116.
[*9*] Eisner, H. S., *The Engineer*, 17 Feb. 1967, pp. 259-260.

John W. Kalis, Jr.[1]

Using Specifications to Avoid Chloride Stress Corrosion Cracking

---

REFERENCE: Kalis, J. W., Jr., **"Using Specifications to Avoid Chloride Stress Corrosion Cracking,"** *Corrosion of Metals Under Thermal Insulation, ASTM STP 880*, W. I. Pollock and J. M. Barnhart, Eds., American Society for Testing and Materials, Philadelphia, 1985, pp. 199–203.

ABSTRACT: Insulated austenitic stainless steel above the critical temperature must be clearly identified to receive specific protection. Specify appropriate insulation and weather barrier material to prevent moisture entry. Proper application of insulation materials is an important part of the specification. Considerations for applying additional protection to austenitic stainless steel are dependent on the operating temperature and critical service of the insulated facility.

KEY WORDS: insulation, specifications, corrosion, stainless steels, protection

It is well recognized that external stress corrosion cracking (ESCC) of austenitic stainless steel is induced by the presence of moisture, chlorides, tensile stress, and temperature in the right combination. Most rain, domestic or plant water, contains chlorides that can be concentrated by evaporation at the stainless steel surface.

Dry insulation does not cause ESCC. However, insulation does provide the vehicle for collecting or directing water to areas of the metal surface where chlorides can be concentrated and cause ESCC.

Therefore, the insulation specification must include a moisture barrier to prevent moisture from entering the insulation system. The specification should specify insulation that can retard the migration of moisture and provide for additional protection, such as a coating or paint system, applied to the surface of critical items. The specification must clearly identify all insulated austenitic stainless steel pipe and equipment along with operating temperature, since ESCC usually takes place above 60°C (140°F).

[1]Project specialist engineer-insulation, E. I. du Pont De Nemours Company, Inc., Wilmington, DE 19898.

The primary objective is to prevent moisture from entering the insulation system. This can be accomplished with some reliability by specifying proper materials and insulating techniques. It should be stated that even with the best insulation materials and installation, all systems require periodic inspection and maintenance to ensure that the weather barrier integrity is intact.

The specification must indicate that the insulation is to be stored and installed dry over a dry surface. Many insulation manufacturers state that moderate amounts of water will not degrade the insulation and that once the process is turned on, the heat will drive the moisture out of the insulation. This is partially true, however, the chlorides in the moisture are left behind, which can cause ESCC. The insulation must be protected from rain until the weather barrier finish is properly applied. This is an area where most insulation contractors have difficulty in complying.

Some insulation absorbs greater amounts of moisture than others. For example, calcium silicate can allow water to migrate through the insulation of an entire pipe system or piece of equipment. This accumulation of moisture can reach saturation without visual detection. A small opening in the weather barrier of a pipe elbow can allow a large amount of water to enter and generate a high concentration of chlorides on the pipe surface. This concentration of chlorides may take place at a remote distance from the source of the moisture entry.

It is preferred to specify an insulation material that retards the spread of moisture. Polyisocyanurate insulation is relatively nonwicking and is preferred for use within its temperature limitations. For higher temperatures, a water-resistant perlite silicate insulation is often used. Nonwicking insulation is used on traced piping because the tracer often cannot generate sufficient heat to drive out water from the system.

For many years the inhibiting action of sodium silicate in the insulation or sodium silicate applied directly to the austenitic stainless steel surface was relied upon for protection. The application of sodium silicate on the metal surface is now specified as minimum protection, even where inhibited insulation is used.

Pressure sensitive tapes are normally specified to secure lightweight insulation to pipe. Many tapes have adhesive that contain chlorides. Chloride cracking on austenitic stainless steel pipe where pressure sensitive tape was applied directly to the pipe surface has been observed. A limited number of chloride-free tapes are available; some of them can be certified to be chloride-free.

No tape of any kind is permitted on the surface of any pipe or equipment, since the monitoring of approved tapes on the job site can be difficult.

Insulation weather barriers must be carefully evaluated for suitability to specific applications. The weather barrier must keep the insulation dry. Aluminum jacketing is specified on pipe and equipment for many heavy industrial applications. The aluminum jacket thickness and application will vary for pipe and equipment. Aluminum jacketing provides an excellent low-cost moisture barrier when installed properly. The insulator is instructed to install all aluminum jacketing in a water shed fashion. All jacket seams are to overlap a minimum of 51 mm (2 in.) on pipe and 76 mm (3 in.) on equipment.

The weather barrier or jacket over pipe insulation is specified as 0.4-mm (0.016 in.) thick aluminum. All fittings, such as tees and elbows, flanges, and valves, are covered with aluminum. Aluminum end caps are used where the pipe insulation terminates at valves and flanges to seal the insulation ends from moisture. The aluminum end caps provide protection when the insulation is removed from the flange or valve for service. In some cases the valve or flange may not be reinsulated immediately, thus exposing the pipe insulation end to moisture.

An aluminum weather barrier is used for most cylindrical equipment such as vessels, columns, or tanks. For equipment up to 1.9 m (42 in.) diameter, 0.6 m (0.024 in.) thick smooth aluminum jacketing with spring tensioned stainless steel bands is normally used. This heavier metal jacket with tensioned bands will hold the jacket firmly against insulated equipment. The bands also prevent moisture from working into the overlapped joints. For equipment larger than 1.9 m (42 in.) in diameter, a pre-engineered panel system is preferred. The panels are normally 2.4 m (8 ft) wide by 1.2 m (4 ft) high. The aluminum jacket is 0.8 mm (0.032 in.) thick and with moisture-resistant polyisocyanurate insulation permanently secured to the jacket. The inside surface of the insulation is faced with an aluminum foil for additional moisture protection. The vertical seam of the panel contains a neoprene gasket for sealing and is interlocked for strength. The panels are secured to the equipment with large spring-tensioned metal bands. Several plant sites have successfully used pre-engineered panels on cylindrical equipment for over ten years. Where large equipment is operating above the allowable temperature for application of the pre-engineered panel, then a separate jacket material is indicated. The deep-corrugated aluminum sheets, 0.6 mm (0.024 in.) thick, provide fewer seams for possible water entry and can be installed with less labor than the rolled smooth aluminum. The corrugations also provide additional strength.

Many insulation system failures on equipment begin with metal flashing improperly designed and applied. High winds can blow away the flashing, exposing the insulation to moisture. Heavy aluminum flashing with a minimum thickness of 0.6 mm (0.024 in.) is normally used to cover projections on equipment. For example, stiffener rings that extend beyond the equipment wall insulation require the fabrication of metal flashing for best protection.

The flashing must be designed for strength and be properly secured to allow for expansion and contraction of the equipment. It must prevent water from collecting and also must shed water. Most insulation contractors are capable of fabricating special flashing requirements, however, it is advisable to provide details in the specification.

It is very difficult to provide a good insulation weather barrier on irregular shapes such as dished heads on vessels. Reinforced mastics have been specified in the past for this application but experience has indicated that frequent maintenance is required to prevent failure. Therefore, this finish is only permitted in areas where the insulation cannot get wet.

A finish of fiber-reinforced polyester (FRP) or epoxy provides excellent resistance to weather, mechanical abuse, and atmospheric corrosion. This finish has no seams and can be sealed around equipment nozzles and other protrusions. Some training is required to properly mix and apply FPR as a finish, but most insulation contractors are familiar with it. FRP is specified for the heads of equipment and irregular shapes where metal forming is not practical. Two component epoxy resins reinforced with synthetic organic fiber open weave cloth have been used successfully. The organic fiber cloth provides greater resilience to expansion than the glass cloth. This is important on large equipment operating at elevated temperatures. This FRP type system is less subject to poor weather conditions during application and has fewer components to mix.

For insulated vertical vessels, install a 4.8 mm (3/16 in.) thick carbon steel angle continuously seal welded around the top of the vessel. The horizontal leg of the angle must be 12.7 mm (1/2 in.) greater than the insulation thickness. This allows the side wall insulation and jacket to slip up under 50.8 mm (2 in.) vertical leg, thus isolating the tank top insulation system from the side wall insulation system. This prevents moisture or spilled product from getting into the side wall insulation, should the top insulation system fail. The water shed angle can be installed on the site should the equipment manufacturer fail to provide it.

All protrusions or penetration through the insulation, such as nozzles, support lugs, and so forth, must be sealed with a bead of good caulking compound. A joint separation greater than 3.2 mm (1/8 in.) is not acceptable for caulking. Caulking is used only where it is impractical to provide metal flashing. Properly installed metal flashing will outlast the best caulking compounds. Only the silicone rubber caulking remains resilient for many years and is resistant to higher temperatures and many chemicals. The pigmented (color) silicone rubber caulking material is specified since it provides greater resistance to higher temperatures and ultraviolet light than the translucent type.

Caulking must be done immediately after the insulation jacket is installed since moisture could enter through the open seams if left uncaulked for a period of time. Caulking must be applied on a clean and dry surface. The bead must never be feather-edged since the life of the seal depends on a uniform material thickness.

Austenitic stainless steel piping and equipment where the insulation can get wet, considered critical to production, or contain hazardous chemicals will be painted before insulation is applied. Areas where the insulation can get wet are defined as outdoors, open structures, or washdown areas or are the presence of fire sprinklers and deluge systems that are periodically tested.

For moderately hot operating temperatures, a two-coat polyamide epoxy system is used. For higher temperatures, a two-coat heat resistant silicone system is specified. The equipment and piping to be painted are identified in the insulation specification. Providing this information will alert the insulation supervisor that the surface must be painted before applying the insulation. Since he is usually not responsible for painting, specific information on surface preparation and painting is normally indicated in the painting specification. Paint provides additional protection to the austenitic stainless steel should moisture-containing chlorides enter the insulation system.

The use of insulation systems for personnel protection on hot pipe and equipment is discouraged. Shielding the hot surfaces with expanded metal will eliminate potential areas for ESCC.

Conclusion

The insulation specification must identify all insulated austenitic stainless steel pipe and equipment and provide precise requirements for protection. It should specify low wicking insulation materials, rigid aluminum jacket weather barrier, aluminum flashing with detailed sketches, and silicone caulking compound. For irregular shapes, such as equipment heads, use a reinforced polyester or epoxy finish. The specification must instruct the insulator to the proper application of the materials. Complying with the above specification should reduce the possibility of ESCC of austenitic stainless steel under insulation.

Herbert A. Moak[1]

Use Inspection as a Means of Reducing Failures Caused By Corrosion Under Wet Insulation

REFERENCE: Moak, H. A., **"Use Inspection as a Means of Reducing Failures Caused By Corrosion Under Wet Insulation,"** *Corrosion of Metals Under Thermal Insulation, ASTM STP 880*, W. I. Pollock and J. M. Barnhart, Eds., American Society for Testing and Materials, Philadelphia, 1985, pp. 204-207.

ABSTRACT: This paper is an overview of what to look for in inspection to prevent the problem of corrosion under wet insulation. To assure quality, do not limit inspection of thermal insulation systems to just the installation. Review the work of the designer as critically as you do that of the installer. An excellent physical installation cannot always cure a poor design. Inspect on a routine basis to look for "tell-tale" signs of pending failures and follow up with prompt repair.

KEY WORDS: inspection, corrosion, insulation, design

To many people, inspection is regarded as something to be done "after the fact." This is unfortunate because many pending failures are created during design and are perpetuated during installation.

There is no question about moisture being the major cause of corrosion, but when it is retained and concealed by thermal insulation, the conditions are much more severe. This is why inspection should start with design. Look for places where nozzles, hangers, thermocouples, conduit, or other penetrations will permit water entry. On equipment, start at the top and work down. On piping, concentrate on penetrations through the top of insulation on horizontal lines and all penetrations on vertical sections.

On vertical equipment, installation of a weathershed ring at the top head is critical (Fig. 1). Protruding stiffener rings are expensive to flash and present a continuous problem with wicking type insulations. Where possible, modify the

[1]Project engineer, E. I. DuPont de Nemours and Company, Inc., Engineering Department, Wilmington, DE 19898.

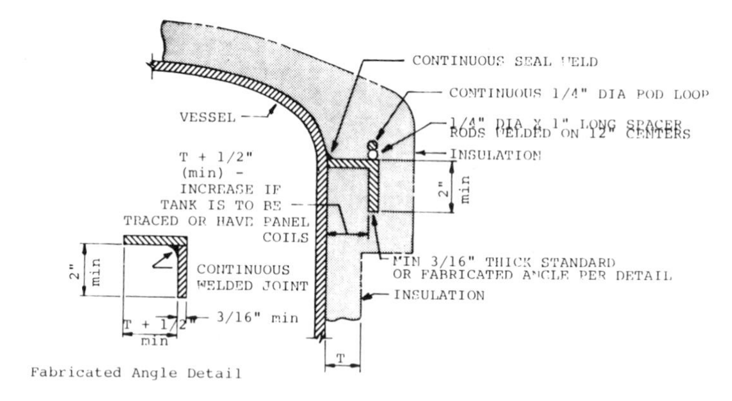

FIG. 1—*Weathershed ring.*

vessel design or insulation thickness or both to permit an uninterrupted exterior finish (Figs. 2 and 3).

Modifications made during design review are less expensive and more effective than modifications made during or after installation. Even the best field application cannot always cure a poor design.

Before applying insulation, make sure that all required attachments have been installed and that where a protective coating has been specified it has been applied over a properly prepared surface.

Another critical inspection function that is frequently overlooked is to review in detail with the supervision responsible for installation (contractor, mainte-

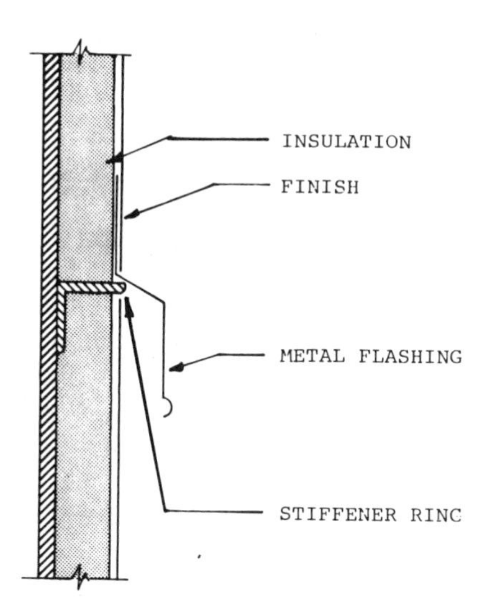

FIG. 2—*Typical installation at stiffener rings.*

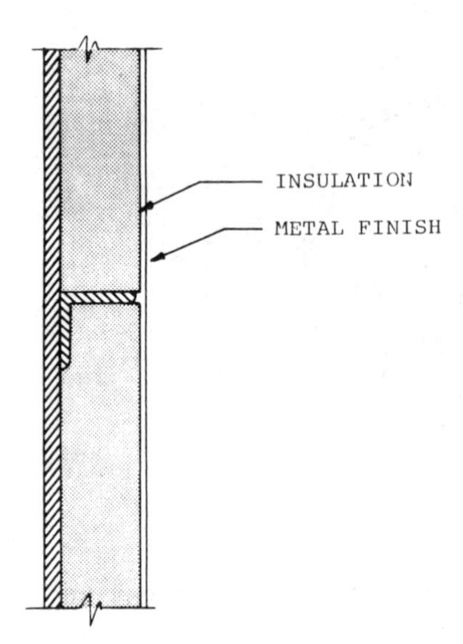

FIG. 3—*Increased insulation thickness to provide uninterrupted finish.*

nance, and so forth) exactly how the work is to be done and the minimum quality level that will be accepted. Do not assume anything. An open discussion can help to avoid later misunderstandings and possible rework.

During installation,

- Make sure installation is dry.
- If insulation is temporarily covered with plastic, be sure it is vented to prevent condensation.
- Use close fitting (6-mm [1/8 in.] maximum opening) metal flashing at all openings.
- Apply silicone caulking at all points of possible moisture entry. Apply a minimum bead of 6-mm (1/4 in.) and do not "feather" edges. "Feathered" edges curl and pull away from the metal.
- Apply all jacketing so that all overlaps shed water.

Inspections following installation must be made frequently on a routine basis by a qualified person. Damage that may seem insignificant to the untrained eye can have serious potential. As an example, a superficial look at caulking or mastic may show that it is in place, however, a closer examination will reveal cracked mastic, exposed reinforcing, and caulking that has pulled away from metal surfaces.

Any gouge or puncture of the finish system can permit water entry. Openings where insulation has been removed for maintenance or modification or both presents the same problem.

Other inspection techniques, such as infrared or ultrasound, can detect "hot spots" or a change in metal thickness, but they do not detect points of water entry and are not a substitute for visual inspection.

Inspection without proper and prompt correction is wasted effort because just knowing there is a problem is not enough.

Conclusion

In summary, inspection is an excellent tool but only when used properly. This means having a qualified owner's representative to

- Review design.
- Inspect daily during construction (installation).
- Periodically and routinely inspect for damage after installation.
- Assure that necessary repairs are made promptly and properly.

This effort is the best insurance against metal corrosion and is the most cost effective.

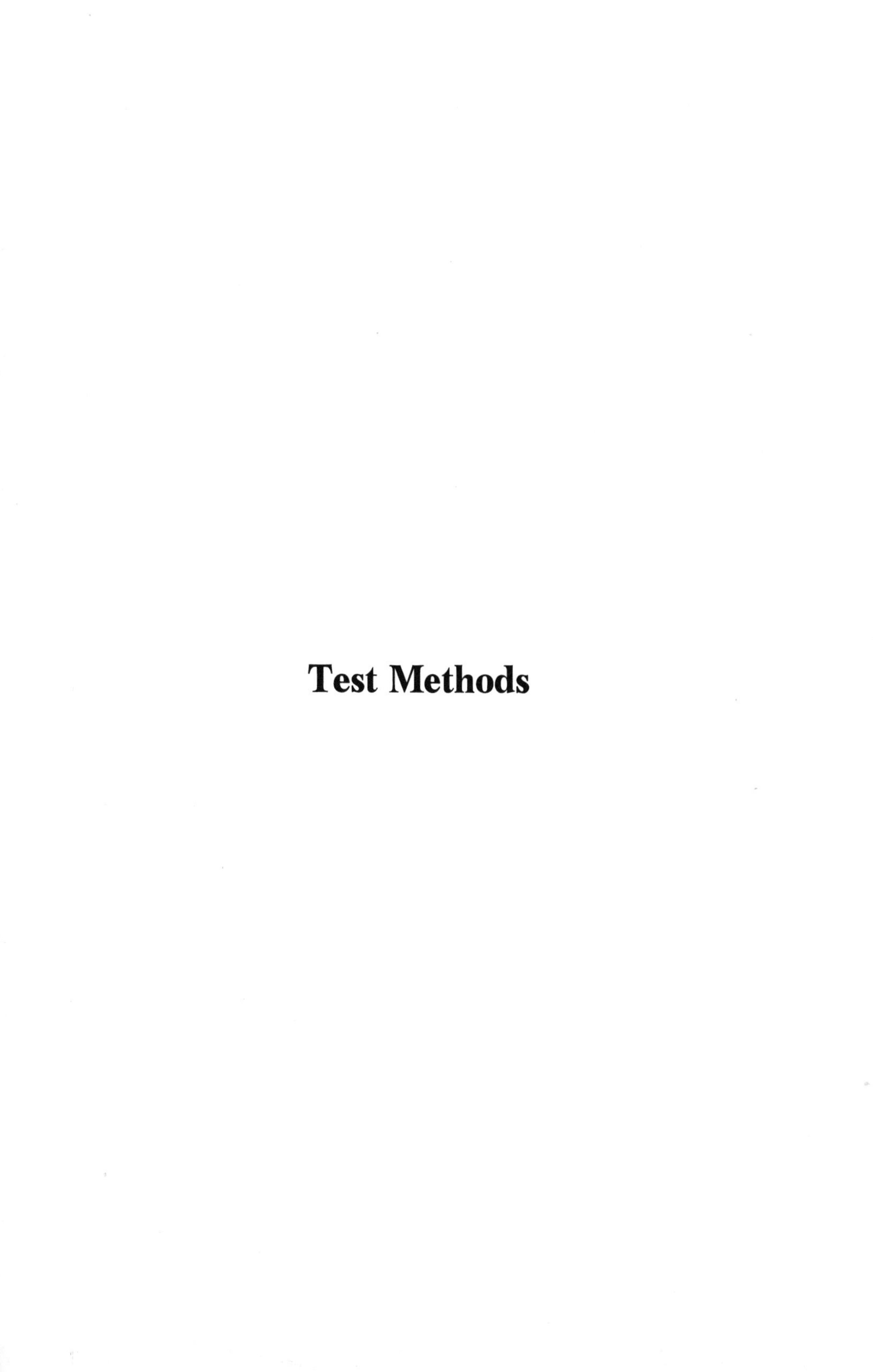

Test Methods

Francis B. Hutto, Jr.,[1] *Ralph G. Tissot, Jr.,*[1]
Thomas E. Whitaker[1]

A New Apparatus and Test Procedure for Running ASTM C 692 Stress Corrosion Cracking Tests

REFERENCE: Hutto, F. B., Jr., Tissot, R. G., Jr., and Whitaker, T. E., **"A New Apparatus and Test Procedure for Running ASTM C 692 Stress Corrosion Cracking Tests,"** *Corrosion of Metals Under Thermal Insulation, ASTM STP 880*, W. I. Pollock and J. M. Barnhart, Eds., American Society for Testing and Materials, Philadelphia, 1985, pp. 211-219.

ABSTRACT: A new experimental technique has been developed for running stress corrosion cracking tests on insulating materials. The new technique uses a steam heated pipe for heating the coupons and uses an "upside down" dripping action rather than a "wicking" action, as in ASTM Evaluating the Influence of Wicking-Type Thermal Insulations on the Stress Corrosion Cracking Tendency of Austenitic Stainless Steel (C 692). In addition to providing a more realistic approach, the new procedure correlates well with the old, simplifies sample coupon preparation and gives added reliability to the test results.

KEY WORDS: stress corrosion, insulation, test apparatus, external stress corrosion cracking

ASTM Evaluating the Influence of Wicking-Type Thermal Insulations on the Stress Corrosion Cracking Tendency of Austenitic Stainless Steel (C 692) has long served as the basic test method for evaluating the influence of wicking-type thermal insulations on the stress corrosion cracking tendency of austenitic stainless steel. The test method itself was a direct outgrowth of the work done at du Pont by Dana and Delong, initially reported in 1956 [*1*]. This test method and other similar procedures are generally referred to as the "Dana Test."

Stated simply, the Dana test calls for the preparation of a precisely stressed U-bend, which is fitted into a block of insulation, which, in turn, is partially

[1]Director of research and development, senior research technician, and senior research engineer, respectively, Pabco Insulation Division, Louisiana-Pacific Corporation, Fruita, CO 81521.

immersed in a dish of distilled water. During the course of the test, the U-bend is heated to near the boiling point, such that water is wicked through the insulation to the U-bend surface where it evaporates. If chlorides are present in the insulation material, they tend to concentrate on the metal surface where external stress corrosion cracking (ESCC) can occur. Based on a large body of data developed by some investigators, it was determined that if ESCC did not take place in 28 days, the material under testing could be considered to be benign insofar as ESCC was concerned in the field.

It was recognized early that even though chlorides contained in insulation materials could contribute to ESCC, chlorides introduced from outside sources were an even more important factor to consider. Ashbaugh reported in 1965 [*2*] that the "frequency of ESCC, as reported, appears to be higher in coastal locations."

A companion test, generally referred to as the "accelerated Dana test," calls for the use of 1500-mg/L chloride solution in place of the distilled water and running the test only 6 days rather than 28 days. This accelerated test was allowed as an alternate pre-production test in Research, Development, Test, and Evaluation (Navy) (RDT) Test Requirements for Thermal Insulating for Use on Austenitic Stainless Steel (M12-1T), which has been abandoned in favor of ASTM C 692.

The accelerated Dana test was testing more than the regular 28-day test. By introducing additional chloride into the system, this test was evaluating the ability of the insulation to inhibit the action of chlorides introduced from outside sources. Several organizations, including du Pont, Hercules, and Ontario Hydro, have recognized this by requiring that an insulating material be capable of passing the accelerated Dana test to qualify for use over austenitic stainless steel. The addition of the accelerated Dana test as an alternate test procedure is currently under ballot in ASTM Committee C 16 on Thermal Insulation.

The objective of this project was to develop an improved test procedure that reduces or eliminates uncontrollable variables in the existing ASTM C 692 test method.

Discussion

Problems with ASTM C 692

In the course of updating ASTM C 692-77, many problems with the current procedure have come to light, particularly as related to the proposed alternate test method utilizing chloride solution rather than distilled water.

Problems with Temperature—ASTM C 692-77 calls for the use of four test coupons, resistance-heated in electrical series with a fifth "dummy" coupon to which is attached a thermocouple for monitoring the uniformity of the coupon temperature control. After establishing, by means of a surface pyrometer that the test coupons are at the proper test temperature (test coupons run at lower temperatures than the dummy, which is in air and has no liquid to wick up), the temperature is controlled by controlling the established offset of the dummy coupon. When each test coupon is instrumented with an attached thermocouple, it becomes clear that the temperature uniformity requirements of ASTM C 692-77 are very difficult to meet (Table 1).

Problems with Evaporation Rate—Water evaporation is impacted by many factors including coupon temperatures, coupon fit to insulation block, tightness of enclosure around test coupon, temperature of test environment, relative humidity of test environment, liquid level in test system, and local boiling point for water. ASTM C 692-77 is in conflict with Military Specification Insulation Material with Special Corrosion, Chloride, and Fluoride Requirements (MIL-I-24244B), which allows the use of C 692-77 test procedures "with certain restrictions." The evaporation rate specified in the MIL-I is not allowed in ASTM C 692–77! With the accelerated Dana test, using 1500-mg/L chloride solution, one might expose different coupons to different amounts of chloride, depending on the evaporation rate and whether makeup is with chloride solution or distilled water.

Electrical Effects—There has been some conjecture about the possible effects of utilizing an electric current for resistance-heating of the coupons.

Testing Difficulty with Certain Materials—As presently constituted, ASTM C 692-77 addresses itself only to wicking-type insulations. If the insulation will not wick, it cannot be tested. There is also some question as to whether or not wicking is a reasonable test mode, since there would be relatively few "real life" situations where the insulation would be partially immersed in a "wicking" mode. Insulations with a water-repellent treatment cannot be tested "as received" but might well wick after exposure to service temperatures. ASTM C 692-77 makes no provision for broadening the applicability of the test by such means.

Searching for a Better Way

In reviewing other test methods and the literature, several observations were made: (1) the hot steam pipe idea of MIL-I-24244B would appear to be a

TABLE 1—*Coupon to coupon temperature variation (typical data).*

ASTM C 692, °C (°F)	Pipe Tester, °C (°F)
96 (205)	97 (206)
111 (231)	98 (208)
108 (226)	98 (208)
94 (201)	96 (204)
103 (217)	95 (207)
157 (315)[a]	98 (208)

[a]Dummy.

better route to temperature uniformity and (2) the dripping approach of Ashbaugh [2] appeared to be a more realistic approach than wicking, perhaps even applicable to nonwicking materials.

The first step in developing a new procedure was to try and duplicate the work of Ashbaugh, using a long, steam-heated pipe that would allow the testing of several coupons at once. The U-bends were tested "upside down" with the insulation on top as shown in Fig. 1. While the coupons could be stress corrosion cracked in this manner, it proved to be next to impossible to devise a "dripper" that would deliver of the order of one drop per minute without slowing down with time. After extensive experimentation, it was found that a multichannel peristaltic pump was the answer. After making several modifications to the steam boiler approach, the test pipe that evolved is shown in Fig. 2.

Since the available peristaltic pump would accommodate 20 channels, the steam pipe was made long enough to accommodate 20 coupons at a time.

Developing a Test Procedure

Having established that the peristaltic pump could deliver 20 channels with an accuracy of ±10%, we set out to determine what drip rate would match the performance of accelerated ASTM C 692-77. After many tries, it was established that a rate of 250 mL/day was the right amount. The apparatus was set up such that each channel of the peristaltic pump had its own graduated 1000-mL reservoir. Thus, as a test proceeded, each channel could be individually monitored to see that the amounts being delivered were within the prescribed limits.

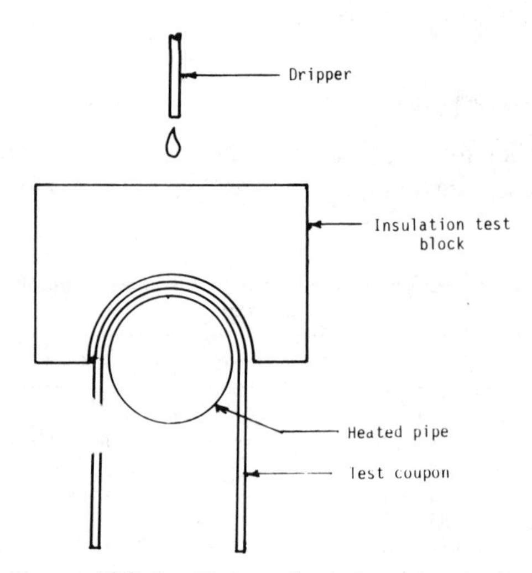

FIG. 1—*Test sample configuration.*

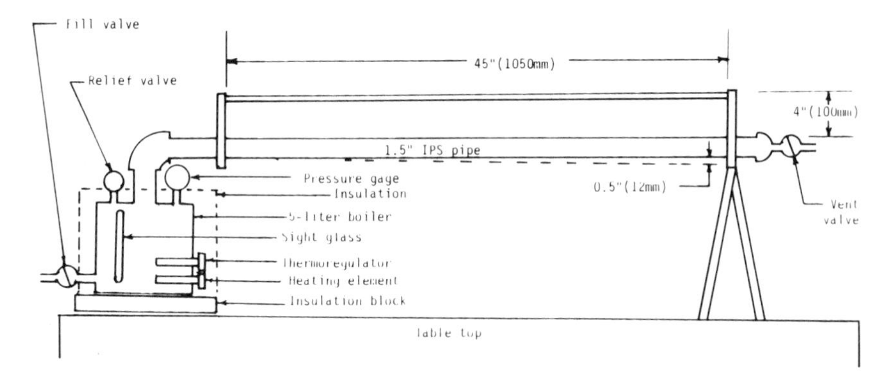

FIG. 2—*Modified ASTM C 692 apparatus.*

It was also judged that a more uniform start would be given the various samples if all were presoaked for 1 h in 1500-mg/L chloride solution by partial immersion in an evaporating dish as shown in Fig. 3.

The following test procedure was established:

1. Prepare test coupons and insulation samples as prescribed in C 692-77. Reserve positions 1 and 20 for "blanks" where the corroding liquid is allowed to drip directly on the coupons. If distilled water is being used, the blank coupons must not crack in order to validate the test. If 1500-mg/L chloride is being used, the blank coupons must crack in order to validate the test.
2. Place coupons side by side on the heating pipe with a dab of heat transfer grease under each. The second and nineteenth coupons must be instrumented by placing a 28 American Wire Gage (AWG) (or finer) thermocouple under the coupon in the heat transfer grease. The test temperature shall be the average of the two thermocouples.
3. Stress coupons in accordance with ASTM C 692-77. Rather than drill holes and stress with through-bolts, it was found that ordinary C-clamps could do an equivalent job with much less effort and extraneous damage to the coupon.

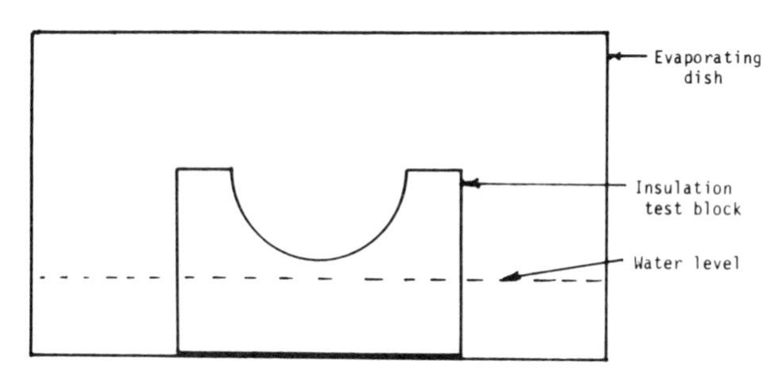

FIG. 3—*Presoaking test sample.*

4. Carefully fit each insulation sample to its coupon by sanding the insulation where necessary. Mark both coupons and samples to match, using a permanent coupon marking system for future reference.

5. Presoak each insulation sample in the test liquid by partial immersion for 1 h as shown in Fig. 3.

6. Turn on pipe heater and bring to temperature (boiling point +6 − 3°C [+10 − 5°F]) while the insulation samples are soaking.

7. Place the soaked insulation samples on their respective coupons and start peristaltic pump.

8. Note starting time (start of pump).

9. With the peristaltic pump, set to deliver 250 mL/day, monitor closely to be sure that all deliveries are within ±10% of this prescribed amount. Positive data points (cracking occurred) where the fluid delivery rate was more than 10% high must be discarded. Negative data points where the fluid delivery rate was more than 10% low must also be discarded. Generally lower rates of flow or no flow may be attributed to crimps or bends of the delivery tubes within the peristaltic cassettes.

10. Monitor temperature as measured by thermocouples on Coupons 2 and 19 and keep the average of the two thermocouples within +6 − 3°C (+10 −5°F) of the boiling point.

11. At the completion of the test period, remove and clean all coupons in accordance with ASTM C 692-77.

12. Examine at ×30 for stress corrosion cracks in accordance with ASTM C 692-77. If in doubt, bend the coupons on a smaller radius over the suspect area. Real cracks will open up.

Validation of Test Procedure

Extensive testing was carried out with two calcium silicate materials, one (Calsil A) which routinely passes the accelerated Dana test, and one (Calsil B) which routinely fails the accelerated Dana test. Both materials routinely pass the regular Dana test, as do most other insulation materials.

When 10 samples of Calsil A and 10 samples of Calsil B were tested simultaneously for six days using 1500-mg/L chloride solution, 10 samples of Calsil A passed (did not crack) while 10 samples of Calsil B failed (cracked). When Calsil A and Calsil B were tested for 28 days with distilled water, both passed.

Test Data

Several materials were tested on both the 28-day and the 6-day tests, and all results were basically in agreement with ASTM C 692-77 (Tables 2 and 3).

TABLE 2—*Test data on various materials run six days at 250 mL/day of 1500-mg/L chloride solution.*

Test	Material	Average Temperature, °C (°F)	Result
1	Calsil A	99 (210)	10 samples, no cracks
	Calsil B		10 samples, all cracked
2	Calsil A	107 (225)	2 samples, no cracks
	Calsil B		2 samples, both cracked
	Calsil C		2 samples, no cracks
	Calsil D		2 samples, both cracked
	Perlite A		2 samples, both cracked
	Perlite B		2 samples, no cracks
	blank		2 samples, both cracked
3	Cellular glass A	96 (205)	2 samples, both cracked
	Perlite A		2 samples, both cracked
	Perlite A heat treated at 260°C (500°F)		2 samples, both cracked
	Perlite B		2 samples, no cracks
	Perlite B heat treated at 260°C (500°F)		2 samples, no cracks
	Fiberglass A		2 samples, both cracked
	Calsil B		2 samples, both cracked
	Calsil C		2 samples, no cracks
	blank		2 samples, both cracked
4	Calsil C	99 (210)	2 samples, no cracks
	Calsil C heat treated at 260°C (500°F)		2 samples, no cracks
	Calsil B		2 samples, both cracked
	Calsil B heat treated at 260°C (500°F)		2 samples, both cracked
	blank		2 samples, both cracked

Noncritical Nature of Coupon Preparation

ASTM C 697-77 would make it appear that coupon preparation and stressing are extremely critical to the reproducibility of the test. To check this point, Calsil A and Calsil B were run side by side with various surface preparations and stressings as shown in Table 3. Based on these data, coupon surface preparation and the degree of post-stressing are very noncritical items. As might be expected, the bending process appears to have provided all of the effective stress necessary to make the test method work. Additional post-stressing appears to be unnecessary. Round-robin testing should quickly establish that less stringent surface preparation is necessary.

Argon Sensitization

A 28-day test was run comparing argon-sensitized coupons with unsensitized coupons. There appeared to be no obvious difference, in that all coupons passed (Table 4). Table 5 shows six-day test results. With additional data, this requirement might also be modified.

TABLE 3—*Effect of surface treatment on ESCC (six-day test at 100°C) (212°F) average temperature).*

Material	Surface Treatment	Result
Calsil A	regular C 692 preparation	no cracks
Calsil B		cracked
Calsil A	bent first, sanded afterward	no cracks
Calsil B		cracked
Calsil A	hand sanded, 80-grit sandpaper	no cracks
Calsil B		cracked
Calsil A	hand cleaned with commercial stainless steel polish	no cracks
Calsil B		cracked
Calsil A	hand cleaned with Ajax® cleanser	no cracks
Calsil B		cracked
Calsil A	not cleaned at all, only degreased	no cracks
Calsil B		cracked
Calsil A	hand cleaned with number 2 coarse steel wool	no cracks
Calsil B		cracked
Calsil A	belt sanded using a worn 80-grit belt	no cracks
Calsil B		cracked
Calsil A	ASTM C 692 preparation, double stressed	no cracks
Calsil B		cracked
Calsil A	ASTM C 692 preparation, no stress beyond original bend	no cracks
Calsil B		cracked

TABLE 4—*Effect of argon sensitization on ESCC (28-day test at 250-mL distilled water/day with coupons at 98°C [208°F]).*

Coupon	Sample	Result
Standard	Calsil A	no cracks
Standard	Calsil A heat treated at 260°C (600°F)	no cracks
Argon-sensitized	Calsil A	no cracks
Argon-sensitized	Calsil A heat treated at 260°C (600°F)	no cracks
Standard	Calsil B	no cracks
Standard	Calsil B heat treated at 260°C (600°F)	no cracks
Argon-sensitized	Calsil B	no cracks
Argon-sensitized	Calsil B heat treated at 260°C (600°F)	no cracks

TABLE 5—*Six-day test using argon-sensitized coupons.*

Sample	Results
Calsil A	no cracks (5 of 5)
Calsil B	all cracked (5 of 5)
Calsil D	all cracked (5 of 5)
Blank	all cracked (2 of 2)

Conclusions

As compared with ASTM C 692-77, the proposed test method:

(1) has better temperature control,
(2) requires less operator attention,
(3) requires less critical sample preparation,
(4) is a more realistic approach to the mode of contamination, and
(5) is more broadly applicable to different types of insulating materials.

References

[1] Dana, A. W., Delong, W. B., *Corrosion*, Vol. 12, No. 7, July 1956, pp. 19-20.
[2] Ashbaugh, W. G., "ESCC of Stainless Steel Under Thermal Insulation," *Materials Protection*, May 1965, pp. 18-23.

Keith G. Sheppard,[1] *Sunil Patel,*[1] *Mukesh Taneja,*[1] *and Rolf Weil*[1]

Comparisons of Several Accelerated Corrosiveness Test Methods for Thermal Insulating Materials

---

REFERENCE: Sheppard, K. G., Patel, S., Taneja, M., and Weil, R., **"Comparisons of Several Accelerated Corrosiveness Test Methods for Thermal Insulating Materials,"** *Corrosion of Metals Under Thermal Insulation, ASTM STP 880*, W. I. Pollock and J. M. Barnhart, Eds., American Society for Testing and Materials, Philadelphia, 1985, pp. 220-230.

ABSTRACT: A study was conducted to provide data on which to base a corrosiveness test for the thermal insulating materials used in residential structures. Several possible test methods were compared. The materials tested included celluloses containing several different fire-retardant additives, glass fiber, rock wool, and a urea formaldehyde foam. Because of their widely differing physical properties, testing was conducted in water leachants made from the insulations. In addition, a test was performed that simulated the condensation conditions that might occur in a residence. It was found that two leachant-based methods could be suitable for accelerated corrosiveness testing of thermal insulation. One method involved determining the corrosion rate of metal coupons immersed in leachant for 14 days at 45°C. The other test was cyclic potentiodynamic voltammetry, which can be completed in only a few hours.

KEY WORDS: thermal insulation, corrosion, corrosion tests, corrosiveness, electrode potentials, anodic polarization, pitting, cellulose, glass fiber, mineral wool, urea formaldehyde foam

The purpose of this study was to provide data upon which to base an accelerated corrosiveness test applicable to different thermal insulation materials, particularly those used in residential structures. The Corrosion Task Group of ASTM Subcommittee C16.31 on Chemical and Physical Properties has been charged with the development of such a test procedure. At the present time, ASTM Specification for Cellulosic Fiber (Wood-Base) Loose-Fill Thermal In-

[1]Assistant professor, graduate assistant, graduate assistant, and professor, respectively, Department of Materials and Metallurgical Engineering, Stevens Institute of Technology, Hoboken, NJ 07030.

sulation (C 739) includes a corrosion test. ASTM C 739 is for loose-fill cellulosic insulation, and the corrosion test procedure is not applicable to other types of insulation. The only other relevant ASTM standard for thermal insulation that includes corrosion testing is ASTM Evaluating the Influence of Wicking-Type Thermal Insulations on the Stress Cracking Tendency of Austenitic Stainless Steel (C 692), but this covers a very specific corrosion condition. The Corrosion Task Group is impeded in its consideration of a new uniform corrosiveness test for thermal insulation by the severe lack of published laboratory and field performance data on the corrosiveness of these materials. The study reported here attempts to provide some of the data necessary to enable rational decisions to be made on possible test procedures.

There are several desired characteristics of a new corrosiveness test. It should be quantitative and be applicable to various insulating materials with their differing physical and chemical properties. It should provide rapid results, so that the test can be applied to quality control during manufacture. The test should be reproducible, and ideally from its results, the behavior of the insulating material under typical field conditions as well as under "worse-case" conditions should be predictable. The ability to predict performance under field conditions is perhaps the area of most controversy in corrosion test development. It is likely to be the least attainable of the desired characteristics because of the many factors other than the inherent corrosiveness of the insulation itself that influence corrosion in the field.

A prerequisite for corrosion is the presence of moisture. Moisture can be provided by condensation, leakage, for example, from a faulty roof, or by absorption caused by hygroscopic components of the insulation material. As an example of the latter, Anderson and Wilkes [1] found that in a moisture absorption test (ASTM C 739), cellulosic insulation containing sulfate fire-retardent additives showed as much as a 40% weight gain in 24 h when exposed to an atmosphere at 90% relative humidity. Weight gains of up to 76% were recorded with longer exposures. The presence of any type of insulation may produce condensation conditions in a wall cavity [2], particularly in the absence of an effective vapor barrier. Thus, the potential for corrosion of metal components may be created solely by the physical presence of insulation and its influence on thermal gradients. Some characteristics of the insulation that may also influence corrosion are the already mentioned moisture content, the permeability of the insulation to moisture, and the nature of the physical contact between insulation and any metal components, particularly the presence of occluded areas that have become anodic because of oxygen deprivation.

Another major factor that can influence corrosion is the presence in an insulation material of chemicals such as fire-retardant additives, binders, and so forth. These ingredients can affect the corrosion rate when conditions exist that allow corrosion to occur. The physical distribution of chemicals within the insulation and the stability of those chemicals over long time periods are also important considerations. In addition to the characteristics of the insulation materials themselves, the environmental variables of temperature, air-flow and humidity, and their variation with time may influence corrosion.

In order to predict corrosion performance in the field and therefore take into account the many variables outlined above, it appears necessary to develop a model containing the various chemical, physical, and environmental parameters. The corrosiveness of the insulation material as determined by any new laboratory test would be but one of the factors in such a model. At present there are insufficient data to develop a suitable model. In particular, field-performance data are virtually nonexistent. The few published field studies [*2,3*] containing data on corrosion were not specifically designed to investigate corrosion. The investigators did not look for metal components in the buildings under study; in fact, generally such locations were avoided. The availability of reliable field data is limited by the cost and difficulty associated with gaining access to metal components in the walls of houses and by problems with obtaining adequate environmental histories at each location. A more fruitful approach is likely to be the simulation of field conditions using a test wall where full control can be exercised over a wide range of conditions.

In view of the above considerations, it is felt that at present, a new corrosiveness test should be limited to assessing the relative aggressive or inhibiting effects of the various ingredients of the insulating materials once conditions favorable for corrosion are created. The possible test methods suitable for a new corrosiveness test are listed in Table 1. Some advantages and disadvantages of

TABLE 1—*Possible corrosiveness test methods.*

1. Visual observation
   1.1 pitting (pass-fail criterion)
   1.2 general corrosion (rating)
   Advantages: no sophisticated equipment needed—easily performed
   Disadvantages: slow, often poorly reproducible, subject to judgement
2. Coupon weight loss
   Advantages: same as 1
   Disadvantages: poor indication of pitting, slow
3. Electrical resistance probes
   (measures resistance change caused by cross-sectional area reduction)
   Advantages: can be used with "dry" insulation
   Disadvantages: poor indication of pitting
4. Linear polarization
   (determines corrosion current (rate) by application of small potential change and extrapolation of assumed linear potential-log current relationship)
   Advantages: rapid quantitative measurement of general corrosion rate as function of exposure time; fairly simple equipment requirements
   Disadvantages: requires "wet" insulation or leachants, poor pitting indication
5. Voltammetry
   (application of increasing and then decreasing anodic potential while current is monitored)
   Advantages: rapid, some pitting indication as well as general corrosion
   Disadvantages: requires "wet" insulation or leachants; relatively sophisticated equipment requirements

each method are mentioned. The techniques were discussed in greater detail in a previous publication [*4*].

Experimental Procedure

The thermal insulation materials studied in this program are listed in Table 2. The choice of those insulation materials to be tested resulted from an initiative by the Ad Hoc Cellulose Industry-Government Agencies Group. This group decided to supply a number of cellulosic insulation batches to be used in various test programs. These batches were therefore included in the present study together with three batches of glass-fiber insulation, one of mineral wool, and one of urea formaldehyde (UF) foam. The cellulose batches chosen by the Ad Hoc Group all contained borax as an additive. As borax is known to be a corrosion inhibitor, it was decided to include an additional cellulose batch (593) containing no borax and known to be relatively corrosive from previous testing [*1*]. Unfortunately, because of the limited quantity of Cellulose 593 available, we were not able to use this material in all the tests.

The test methods (refer to Table 1) compared in this study included weight loss and appearance of steel coupons. The corrosion rate of steel coupons was also determined by the polarization-resistance method. These tests will be referred to as the "leachant-coupon" tests. In addition, values of the limiting current density together with observations of the presence or absence of a positive hysteresis loop were determined by potentiodynamic cyclic voltammetry (Fig. 1) on steel. This testing will be referred to as voltammetry. It should be noted that limiting current density determined by voltammetry (that is, the value at which the current limits as the anodic potential applied to the specimen is increased), is an arbitrary measure in that its value is subject to mass transport conditions.

TABLE 2—*Insulating materials used in test program.*

Insulation	Chemical Additive
Cellulose 1	1 part borax (5 mole); 2 parts boric acid, 25% chemical content
Cellulose 2	2 parts borax; 1 part boric acid, 25% chemical content
Cellulose 3	1 part borax; 1 part boric acid; 1 part aluminum trihydrate, 25% chemical content
Cellulose 4	1 part borax; 4 parts ammonium sulfate, 30% chemical content
Cellulose 6	2 parts borax; 2 parts boric acid; 1 part aluminum sulfate, 25% chemical content
Glass Fiber A	no binder
Glass Fiber B	with binder
Glass Fiber C	with binder[a]
Rock wool	...
Urea-formaldehyde foam	...
Cellulose 593	aluminum sulfate, 17% chemical content

[a] Similar density and fiber size to B, but different binder.

However, it was felt that the limiting current density provided a means for making a relatively simple assessment of corrosiveness under standardized conditions, and it had been found in earlier work [4] to correlate well, at least for cellulosic materials, to the relative corrosiveness determined by embedding metal coupons in wet insulation. Figure 1 shows examples of typical voltammetry curves obtained in our earlier work with cellulosic insulation containing different fire-retardant additives [4]. The limiting current density is indicated on Curve a. Curve c shows the positive hysteresis loop on reversing the anodic potential scan that is typically obtained for pitting-type corrosion.

The test methods outlined above were compared using steel specimens immersed in leachants (that is, water extracts of soluble components) made from the different insulation samples. The reason for using leachants rather than the insulation materials themselves is that some types of insulation are relatively hydrophobic, and it is difficult to achieve the conductive path between electrodes necessary for application of the electrochemical testing methods. Previous tests [4] had shown that leachants from cellulosic insulation produced values of relative corrosiveness similar to those obtained using the wet insulation materials. It was desirable to use an existing standard leaching method if possible. Therefore the leaching procedure used in this study was essentially that specified in ASTM Chemical Analysis of Thermal Insulation Materials for Leachable, Chloride, Silicate, and Sodium Ions (C 871), however the amount of each type of insulation leached was based on the ratios of the applied densities of the materials obtained from manufacturers' data. This approach was considered more appropriate than using equal weights, because of the widely differing densities of the materials being compared. The weight of each mate-

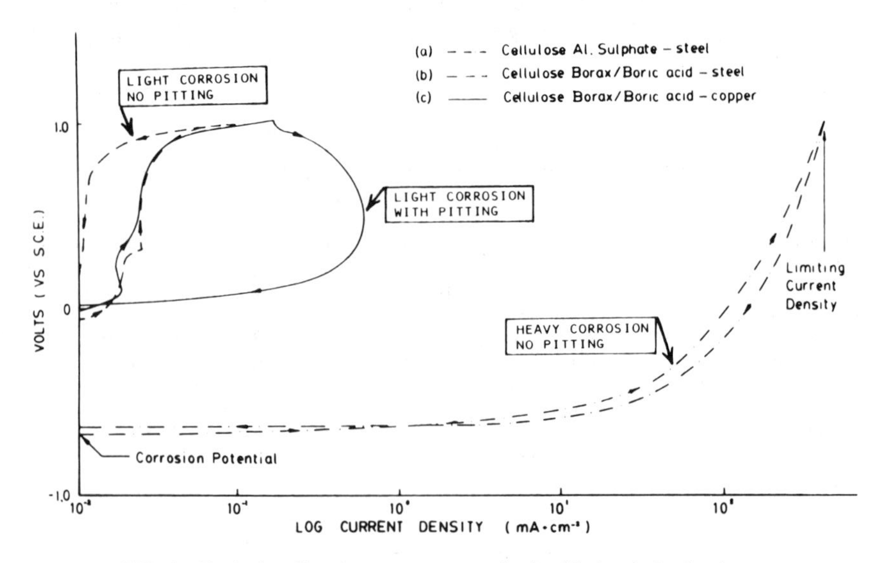

FIG. 1—*Typical cyclic voltammetry curves obtained in insulation leachants.*

rial leached was equivalent to that required to fill the 102 by 102 by 102 mm (4 by 4 by 4 in.) volume used in the condensation test (to be described later).

For the leachant-coupon tests, the Society of Automotive Engineers (SAE) 1010 (Unified Numbering System [UNS] 10100) coupons 51 by 51 by 0.076 mm (2 by 2 by 0.003 in.) were prepared and cleaned according to the ASTM C 739 procedure. Two coupons were fully immersed in leachant contained in a polypropylene dish with a fitted lid. Duplicate dishes were prepared for each insulation batch. Duplicate blanks were prepared using distilled water in place of a leachant. The leachant level was marked on the outside of each dish, and the dishes maintained in an environmental chamber at 45°C and 95% relative humidity for 14 days. The dishes were inspected regularly, and distilled water used where necessary to maintain the liquid level in each dish. At the completion of the test, the coupons were visually assessed for corrosion and then cleaned according to the ASTM Recommended Practice for Preparing, Cleaning, and Evaluating Corrosion Test Specimens (G 1) and their corrosion rates determined from weight loss measurements. Corrosion rate measurements obtained by the polarization-resistance method were carried out using a similar experimental arrangement except that three coupons were used in each dish. Insulated wires were attached to each coupon, and then the soldered joints were lacquered. The three coupons became the anode, cathode, and reference electrodes when immersed in leachant and were polarized using a Petrolite M-1010 instrument (Rohrbach Industries). This instrument operates on the polarization-resistance principle and gives a direct readout of corrosion rate. Corrosion rate readings were monitored throughout the 14-day test period.

Specimens for voltammetry consisted of SAE 1010 steel mounted in epoxy resin and abraded to a 600-grit finish. The area of steel exposed was 1 cm^2. Each specimen was carefully inspected to ensure that no crevice existed between the steel and its epoxy mount. Cyclic voltammetry curves were obtained using a Princeton Applied Research Model 175 Programmer controlling a Model 173 Potentiostat such that the specimen potential was scanned anodically at 1 mV/s in stagnant deaerated leachant. The cell current necessary to maintain the applied potential was plotted on a logarithmic scale versus the applied potential. The scan was reversed at 1.0 V versus a saturated calomel reference electrode. Specimens were allowed to establish a constant corrosion potential in the leachant before commencement of the anodic scan. From the recorded curves, values of corrosion potential and limiting current density were obtained, and the presence or absence of a positive hysteresis loop, indicative of pitting, was noted. A microscopic evaluation of each specimen was made after the test. As with the leachant-coupon test, distilled water was used to run a blank experiment.

In addition to the leachant-based tests described above, the same batches of insulation were tested by placing them dry into 102- by 102- by 102-mm (4- by 4- by 4-in.) compartments within a Plexiglas® box, the bottom of which was a lacquered, water-cooled copper plate. This allowed a 10°C temperature gradient to be established through the insulation specimens when the box was located in an environmental chamber held at 45°C and 75% relative humidity for 65 days. SAE 1010 steel coupons prepared as previously described were embedded horizontally at one-quarter, half, and three-quarters of the depth of each insulation specimen. The dew point would have occurred at approximately half the depth of the insulation under the conditions described with a 10°C temperature gradient, creating the probability of condensation in the lower half of each insulation specimen. On completion of the test, corrosion of the steel was assessed by visual observation and by measurement of coupon weight loss after cleaning by the ASTM G 1 method. This test will be referred to as the "condensation" test. The same experiment was also carried out with no temperature gradient applied through the insulation.

Results and Discussion

The test involving steel coupons embedded in initially dry insulation and exposed for 65 days at 45°C and 75% relative humidity without a temperature gradient through the insulation, resulted in negligible corrosion of the steel. No corrosion was obtained greater than 0.03 mm per year as determined by coupon weight loss. This result suggests that in the absence of moisture from condensation or a leak, there is little likelihood of corrosion occurring in the presence of the batches of insulation tested in the present program. In contrast, the results obtained when a 10°C temperature gradient was set up through the insulation are recorded in Table 3. Corrosion is measured by coupon weight loss and by visual assessment. Only data for the bottom coupon in each insulation specimen are given, that is, the coupon nearest the cooled plate and therefore within the condensation zone. These bottom coupons experienced significantly greater corrosion, in specimens where corrosion occurred, than the two other coupons in each insulation specimen. Corrosion rates for the upper coupons, that is, located above the condensation zone, were all less than 0.03 mm per year. Coupons located at the mid-depth of the insulation exhibited corrosion rates intermediate to those obtained from the upper and lower coupons for all insulation specimens. The data in Table 3 indicate that the coupons in glass fiber and rock wool insulation exhibited little corrosion. Of the cellulosic insulations, Celluloses 1 and 2 containing solely boron fire-retardant additives, produced the least corrosion. Celluloses 4, 6, and 593, containing sulfate fire-retardants, produced relatively severe corrosion.

Data from the leachant-coupon and voltammetry tests are also given in Table 3 to allow comparison. Values of leachant pH are recorded. Only the average corrosion rates for the 14-day test period are given, although corrosion rates were recorded on a daily basis by the polarization resistance method. The presence of a positive hysteresis loop in the voltammetry curves is indicated. The visual appearance ratings of corrosion in each test are tabulated; the rating scheme is in Footnote *b* of Table 3.

Table 4 shows the results of the various tests summarized in terms of corro-

TABLE 3—*Corrosiveness test results for steel.*

	Type of Insulation											
	Cellulosic						Glass Fiber					
	1	2	3	4	6	593	A	B	C	Rock Wool	UF Foam	Distilled Water
Corrosion Rates												
Condensation test from weight loss, mm/y	0.08	0.08	0.13	>0.2	0.1	>0.2	0.03	<0.03	<0.03	<0.03	0.08	...
Leachant coupon tests[a]												
Polarization resistance, mm/y	0.03	<0.03	<0.03	0.11	0.06	...	<0.03	<0.03	<0.03	<0.03	0.05	0.03
From weight loss, mm/y	0.03	<0.03	<0.03	0.2	0.10	...	<0.03	0.03	<0.03	0.03	0.08	0.03
Leachant voltammetry												
Limiting current density, mA/cm^2	0.04	0.04	0.05	0.1	0.07	80	0.03	0.03	0.03	0.03	0.08	0.04
Corrosion potential (−ve), mV versus SCE	254	156	248	665	266	659	177	642	356	157	483	149
Positive hysteresis loop present	yes	no	no	yes	yes	no	yes	yes	yes	yes	yes	no
Leachant, pH	7.6	8.2	8.2	7.7	7.4	7.7	9.0	8.3	9.2	8.0	2.9	6.4
Appearance[b]												
Condensation test	1	2	2	3	3	3	0	0	1	1	2	...
Leachant coupon test	1	2	2	3	3	...	2	2	2	3	2	0
Voltammetry	1	0	0	3	1	3	2	2	2	1	2	0

[a]Typical amounts leached into 1 L of water. 128 g/L cellulose, 25 g/L glass fiber, and 102 g/L rock wool.
[b]0 = no pitting, 1 = slight pitting, 2 = perforations, and 3 = partial or complete coupon corrosion.

TABLE 4—*Corrosiveness rating versus distilled water.*

	Type of Insulation										
	Cellulose						Glass Fiber			Rock Wool	Foam
	1	2	3	4	6	593	A	B	C		
Corrosion rates grouping											
Condensation tests[a]	A	A	A	A	A	A	B	B	B	B	A
Leachant coupon tests[b]											
Polarization resistance	B	B	B	A	A	...	B	B	B	B	A
Weight loss	B	B	B	A	A	...	B	B	B	B	A
Voltammetry[b]	B	B	A	A	A	A	B	B	B	B	A

[a] A represents a corrosion rate > 0.03 mm/y and B represents a corrosion rate ≤ 0.03 mm/y.

[b] A represents corrosion rates greater than in distilled water and B represents corrosion rates equal to or less than in distilled water.

siveness compared to distilled water. The letter A was assigned where corrosion was greater than and B when it was equal to or less than that produced by distilled water. The condensation test results cannot be compared with distilled water in the same way. However, in order to make a comparison to the leachant test results, an arbitrary index was assigned to the condensation test results such that A was given to corrosion rates greater than 0.03 mm per year and B to corrosion rates equal to or less than 0.03 mm per year.

Reasonable agreement is seen in Table 3 between average corrosion rates obtained from coupon weight loss and by the polarization-resistance method for steel in leachants of the different insulation materials. Comparing corrosion rates obtained in the leachant tests with those obtained in the condensation test, one also finds reasonable agreement. Glass fiber and rock wool specimens produced very low corrosion rates in both types of tests. Celluloses 4 and 6 containing sulfates were seen to be corrosive in both types of test. Celluloses 1, 2, and 3 had higher corrosion rates in the condensation test than in the leachant tests.

The limiting current densities obtained from voltammetry show good correlation with the corrosion rate data of the other tests, particularly the leachant-based tests. Corrosion potentials measured in the voltammetry, however, do not show such good correlation. Certainly, Celluloses 4 and 593, the most corrosive in the other tests, had the most active corrosion potentials. Cellulose 2, Glass Fiber A, and Rock wool had the most noble corrosion potentials and very low corrosion rates. However, among the other insulation batches, a correlation between corrosion potential and corrosion rate is not apparent. A positive hysteresis loop was obtained in the voltammetry curves for all specimens except Celluloses 2, 3, and 593 and the distilled water. Microscopic observations of the steel specimens after the tests confirmed the presence of pitting corrosion in all cases where a positive hysteresis loop was obtained.

No direct correlation between leachant pH and corrosion rate was observed. This is in agreement with our previous findings [*4*] and that of others [*1,5*]. The appearance ratings in Table 3 shows only fair agreement when comparing the different test methods for a particular insulation. The visual assessment is inherently subjective and likely to be unreliable as a standard test. For each insulation batch, the A and B ratings of Table 4, comparing corrosion rates to those in distilled water, are found to agree for all the testing methods. Exceptions are Celluloses 1, 2, and 3 for which greater corrosion was produced in the condensation test than in the leachant tests. Such a rating against distilled water might provide a pass/fail assessment of corrosiveness in any new testing standard.

Conclusion

It appears that a leachant-based test is capable of indicating the relative corrosiveness of different types of thermal insulation. Corrosiveness in this sense is limited to a measurement of the effect of water-soluble components of the insulation on the corrosion rate, provided that environmental conditions exist that are conducive to corrosion. It has been shown that the results obtained in the leachant-based tests compare well with those obtained in a simulation of the type of condensation conditions that might occur in service, although the conditions were somewhat severe in this case. Further simulation and field data are needed.

Coupon weight losses determined from the leachant-coupon test or the limiting current density obtained from voltammetry appear to be equally effective tests. Voltammetry has the advantage of a short test period (a few hours) and is therefore useful for quality control. In addition, voltammetry provides an indication of the pitting tendency of the leachable chemicals.

Based on the results reported in this paper, the ASTM C16.31 Corrosion Task Group is conducting a round-robin program to evaluate a leachant-coupon type corrosiveness test for thermal insulation. It plans to conduct a round-robin for a voltammetry-based corrosiveness test in the near future.

References

[*1*] Anderson, R. W. and Wilkes, P., "Survey of Cellulosic Insulation Materials," Technical Report ERDA 77-23, Energy Research and Development Administration, Washington, DC, 1977.

[*2*] "A Field Study of Moisture Damage in Walls Insulated Without a Vapor Barrier," Technical Report ORNL/Sub 78/977261/1, Oak Ridge National Laboratory, Oak Ridge, TN, May 1980.

[*3*] "Minesota Retrofit Insulation In-Situ Program," Technical Report HCP/W 2843-91, U.S. Department of Energy, Washington, DC, 1978.

[*4*] Sheppard, K. G. and Weil, R., *Thermal Insulation, Materials and Systems for Energy Conservation in the '80s, STP 789*, F. A. Govan, D. M. Greason, and J. D. McAllister, Eds., American Society for Testing and Materials, Philadelphia, 1982, pp. 132–144.

[*5*] Shen, K. K., *Journal of Thermal Insulation*, Vol. 3, Jan. 1980, p. 190.

Author Index

Subject Index

D

E

J

L

M